Birkhäuser

Trends in Mathematics

Research Perspectives Ghent Analysis and PDE Center

Volume 1

Series Editor
Michael Ruzhansky, Department of Mathematics, Ghent University, Gent, Belgium

Research Perspectives Ghent Analysis and PDE Center is a book series devoted to the publication of extended abstracts of seminars, conferences, workshops, and other scientific events related to the Ghent Analysis and PDE Center. The extended abstracts are published in the subseries Research Perspectives Ghent Analysis and PDE Center within the book series Trends in Mathematics. All contributions undergo a peer-review process to meet the highest standard of scientific literature.

Volumes in the subseries will include a collection of revised written versions of the communications or short research announcements or summaries, grouped by events or by topics. Contributing authors to the extended abstracts volumes remain free to use their own material as in these publications for other purposes (for example a revised and enlarged paper) without prior consent from the publisher, provided it is not identical in form and content with the original publication and provided the original source is appropriately credited.

Michael Ruzhansky • Berikbol Torebek
Editors

Extended Abstracts MWCAPDE 2023

Methusalem Workshop on Classical Analysis and Partial Differential Equations

Birkhäuser

Editors
Michael Ruzhansky
Ghent Analysis and PDE Center,
Department of Mathematics: Analysis,
Logic and Discrete Mathematics
Ghent University
Ghent, Belgium

School of Mathematical Sciences,
Queen Mary University of London,
London, UK

Berikbol Torebek
Institute of Mathematics and Mathematical
Modeling
Kazakhstan

Ghent Analysis and PDE Center
Ghent, Belgium

ISSN 2297-0215 ISSN 2297-024X (electronic)
Trends in Mathematics
ISSN 2948-1724 ISSN 2948-1732 (electronic)
Research Perspectives Ghent Analysisand PDE Center
ISBN 978-3-031-41667-5 ISBN 978-3-031-41665-1 (eBook)
https://doi.org/10.1007/978-3-031-41665-1

This book is published under the imprint Birkhäuser, www.birkhauser-science.com by the registered company Springer Nature Switzerland AG
The registered company address is: Gewerbestrasse 11, 6330 Cham, Switzerland

Preface

This volume is the first volume in the newly established Birkhäuser series Research Perspectives Ghent Analysis and PDE Center. This is a book series devoted to the publication of extended abstracts of seminars, conferences, workshops, and other scientific events related to the Ghent Analysis and PDE Center. Volumes in this subseries include a collection of revised written versions of the communications or short research announcements or summaries, grouped by events or by topics.

Ghent Analysis and PDE Center is a research group in mathematics that focuses on different areas of analysis as well as the study of partial differential equations (PDEs) and their applications. The details and activities of the center can be assessed at the following website:

https://analysis-pde.org

In addition to the research activities, the Ghent Analysis group is also involved in training the next generation of mathematicians. They organize seminars, workshops, and conferences to share their research findings and collaborate with other researchers in the field. The center brings together researchers from different disciplines to collaborate on projects related to analysis and partial differential equations, and provides a platform for sharing knowledge and expertise.

The activities of the center have been supported by grants from the FWO as well as by the Methusalem program of the Ghent University.

The present volume contains the contributions of the participants of the Methusalem Workshop on Classical Analysis and PDEs which was held at the Ghent University, Belgium, from 27 February to 1 March 2023. The workshop was organized by the Ghent Analysis and PDE Center with the financial support of Methusalem program of the Ghent University Special Research Fund (BOF) (Grant number 01M01021).

The aim of the workshop was to exchange the recent progress and ideas in the field of analysis in PDEs. The workshop was held in a hybrid format with broadcast via Zoom, but all speakers spoke in person. The scientific event was attended by about 70 participants from different countries such as Belgium, Brazil, Bulgaria, China, Ethiopia, France, India, Israel, Kazakhstan, the Netherlands, Turkey, Uzbekistan, and others.

The scientific program of workshop consisted of plenary presentations and short presentations. The following speakers made plenary presentations:

- Ravshan Ashurov, Doctor of Physical and Mathematical Sciences, Professor, Institute of Mathematics, Uzbekistan
- Anatoly Golberg, PhD, Associate Professor, Holon Institute of Technology, Israel
- Tynysbek Kalmenov, Doctor of Physical and Mathematical Sciences, Professor, Academician of the National Academy of Sciences of the Republic of Kazakhstan, Institute of Mathematics and Mathematical Modeling, Kazakhstan
- Baltabek Kanguzhin, Doctor of Physical and Mathematical Sciences, Professor, Al-Farabi Kazakh National University, Kazakhstan
- Roman Novikov, Doctor of Physical and Mathematical Sciences, Centre de Mathématiques Appliquées, École Polytechnique, France
- Erlan Nursultanov, Doctor of Physical and Mathematical Sciences, Professor, Lomonosov Moscow State University, Kazakhstan Branch, Kazakhstan
- Makhmud Sadybekov, Doctor of Physical and Mathematical Sciences, Professor, Institute of Mathematics and Mathematical Modeling, Kazakhstan
- Nedyu Popivanov, Doctor of Sciences, Professor, University of Sofia, Bulgaria
- Marian Slodicka, Doctor of Sciences, Professor, Ghent University, Belgium
- Nazerke Tleukhanova, Doctor of Physical and Mathematical Sciences, Professor, Gumilyov Eurasian National University, Kazakhstan

The following speakers presented short presentations:

- Irfan Ali, PhD Student, Ghent University, Belgium
- Santiago Gomez Cobos, PhD Student, Ghent University, Belgium
- Mulugeta Alemayehu Dagnaw, PhD, Injibara University, Ethiopia
- Jozefien D'haeseleer, PhD, Ghent University, Belgium
- Arne Hendrickx, PhD Student, Ghent University, Belgium
- Junqing Huang, PhD Student, Ghent University, Belgium
- Nurbek Kakharman, PhD Student, Institute of Mathematics and Mathematical Modeling, Kazakhstan
- Bakytbek Koshanov, Doctor of Physical and Mathematical Sciences, Professor, Al-Farabi Kazakh National University, Kazakhstan
- Mostafa Meliani, PhD Student, Radboud university, the Netherlands
- Joel E. Restrepo, PhD, Ghent University, Belgium
- Serikbol Shaimardan, PhD, Ghent University, Belgium

- Abhilash Tushir, PhD Student, Indian Institute of Technology, Delhi, India
- Karel Van Bockstal, PhD, Ghent University, Belgium
- Juan Pablo Velasquez-Rodriguez, PhD Student, Ghent University, Belgium
- Xuechao Wang, PhD Student, Ghent University, Belgium
- Rong Zhang, PhD Student, Nanjing Normal University, China

This volume contains selected extended abstracts from the speakers of the above workshop:

In addition, this volume also includes extended abstracts from the following speakers who made presentations at an associated activity closer to the workshop date

- Arran Fernandez, PhD, Eastern Mediterranean University, Turkey
- Brian Grajales, PhD, UNICAMP Universidade Estadual de Campinas, Brazil

We have separated the topics of presentations in two rough groups: theory of functional analysis, operator theory and applications, as well as classical and non-classical problems of PDEs. The conference included two talks on topics related to computer science, which we included in the first group, as certain applications of mathematical analysis and modeling.

In conclusion, we would like to thank all the authors who have contributed to this volume.

This volume is the first volume of the Research Perspectives Ghent Analysis and PDE Center series.

Birkhäuser Series
Research Perspectives
Ghent Analysis and PDE Center

It is expected that this series will be continued in the near future with the following forthcoming volumes:

Volume title: Extended Abstracts 2021/2022
Volume subtitle: Methusalem Lectures
Volume editors: Duvan Cardona, Joel Restrepo, Michael Ruzhansky;

Volume title: Extended Abstracts 2021/2022
Volume subtitle: Ghent Analysis and PDE Seminar
Volume editors: Michael Ruzhansky, Karel Van Bockstal;

Volume title: Women in Analysis and PDE
Volume editors: Marianna Chatzakou, Michael Ruzhansky, Diana Stoeva;

Volume title: Analysis and PDE in Latin America
Volume subtitle: ICMAM 2022 Latin America
Volume editors: Duvan Cardona, Brian Grajales;

Volume title: Extended Abstracts 2022
Volume subtitle: Analysis and Applied Mathematics
Volume editors: Allaberen Ashyralyev, Michael Ruzhansky, Makhmud Sadybekov.

Ghent, Belgium

Michael Ruzhansky
Berikbol Torebek

Contents

Part I
Theory of Functional Analysis, Operator Theory, and Applications

Mappings of Finite Area Distortion: Hausdorff Measures

Elena Afanas'eva and Anatoly Golberg

Abstract In this paper we continue studying homeomorphisms of finite area distortion in the frame of Hausdorff measures.

2000 Mathematics Subject Classification 30C65, 28A78, 26B35

1 Introduction

In this short paper, we continue studying classes of mappings employing purely metric approach. In [3–5], we established various properties of finitely bi-Lipschitz homeomorphisms, mappings of finite metric and area distortions between Riemannian and Finsler manifolds. Some of these results have been applied to the boundary correspondence problems studied earlier by the first author in [1, 2].

Here we concentrate on establishing various Hausdorff measure bounds for the class of homeomorphisms of finite area distortion introduced in [19]; see also [21]. Our results including at most sketches of proofs will be given in the Euclidean setting only. An extended version in the frame of Riemannian manifolds and containing explicit proofs and descriptions will be presented in our forthcoming paper. Note that by their features, the bounds will be close to estimates obtained in [11, 14–16, 28]. However, we involve a refine kind of estimating which relies on directional dilatations introduced in [13] and [17].

From geometric point of view, any Hausdorff measure bounds can be treated as rectifiability problems [22]. It is well-known fact that even in the two-dimensional case, a quasicircle (a quasiconformal image of circle) can be a snowflake curve and has the Hausdorff dimension which is strongly greater than one; see, e.g., [23]

E. Afanas'eva
Institute of Applied Mathematics and Mechanics of the NAS of Ukraine, Slavyansk, Ukraine

A. Golberg (✉)
Department of Mathematics, Holon Institute of Technology, Holon, Israel
e-mail: golberga@hit.ac.il

M. Ruzhansky, B. Torebek (eds.), *Extended Abstracts MWCAPDE 2023*,
Research Perspectives Ghent Analysis and PDE Center 1,
https://doi.org/10.1007/978-3-031-41665-1_1

and the references therein. In higher dimensions, the situation becomes much more complicated. For Lipschitz mappings, we recall the following result given in [23], which ensures that $\mathcal{H}^{\alpha} f(E) \leq L^{\alpha} \cdot \mathcal{H}^{\alpha} E$, where $\mathcal{H}^{\alpha} E$ stands for the α-dimensional Hausdorff measure of a set $E \subset \mathbb{R}^n$, and $f : E \to \mathbb{R}^n$ is a mapping such that

$$\limsup_{x \to y, x \neq y} \frac{|f(x) - f(y)|}{|x - y|} \leq L < \infty \qquad \forall x \in E. \tag{1.1}$$

A simple example [23] shows that the constant L cannot be dropped even for the case of compact set E. Indeed, consider a mapping $f(x) = (x, g(x))$ in $\mathbb{R}^2$, where $g(x)$ is piecewise linear on $[0, 1]$ and satisfies

$$g(0) = 0, \quad g\left(\frac{1}{n}\right) = \frac{1}{n}, \quad g\left(\frac{1}{2}\left(\frac{1}{n} + \frac{1}{n+1}\right)\right) = 0.$$

Clearly, $\mathcal{H}^1 f[0, 1] = \infty$, whereas $\mathcal{H}^1[0, 1] = 1$. Note also that the limsup in (1.1) does not bring any weakness for the standard Lipschitz mappings where $|f(x) - f(y)| \leq \mathcal{L}|x - y|$ for any x, y, since the relationship between L and $\mathcal{L}$ is quantitative; see, e.g., [6].

2 Main Notions and Axillary Results

Let a mapping $f : D \subset \mathbb{R}^n \to \mathbb{R}^n$, $n \geq 2$, be differentiable at a point $x \in D$, with the non-vanishing Jacobian $J_f(x)$. Then the standard dilatation coefficients (*inner and outer*) of f are defined at x by

$$K_I(x, f) = \frac{J_f(x)}{l^n(x, f)}, \qquad K_O(x, f) = \frac{||f'(x)||^n}{J_f(x)},$$

where $||f'(x)|| = \max_{|h|=1} |f'(x)h|$, $l(x, f) = \min_{|h|=1} |f'(x)h|$, cf. [29].

Fix now an arbitrary point $x_0 \in \mathbb{R}^n$ and denote by $\partial_h f(x)$ the derivative of f at x in the direction h. The *angular dilatation* $D(x, x_0, f)$ of f at a point $x \in D$ with respect to x_0, $x \neq x_0$, is defined by

$$D(x, x_0, f) = \frac{J(x, f)}{\ell^n(x, x_0, f)}, \qquad \ell(x, x_0, f) = \min_{|h|=1} \frac{|\partial_h f(x)|}{|\langle h, \frac{x - x_0}{|x - x_0|}\rangle|}, \tag{2.1}$$

see [17]. The dilatation $D(x, x_0, f)$ is a measurable function in D, and satisfies the following bounds

$$\frac{1}{K_O(x, f)} \leq D(x, x_0, f) \leq K_I(x, f).$$

Replacing in (2.1) the exponent n with a real parameter p, $p \geq 1$, we define the p-angular dilation $D_p(x, x_0, f)$, cf. [14]. The calculation of directional derivatives is more complicated than the standard ones. However, these quantities provide much sharper estimates; see [13, 14, 17, 20].

For example, for a "slow rotation" of the unit ball $\mathbb{B}^n(0, 1)$ in the plane $x_3 = \ldots = x_n = 0$ $(z = x_1 + ix_2)$ defined by

$$g(x) = \left(ze^{i\sqrt{-\log|z|}}, x_3, \ldots, x_n\right), \quad g(0) = 0, \quad |x| < 1,$$

the inner and outer dilatations coincide

$$K_O(x, g) = K_I(x, g) = \left(\frac{\sqrt{16\log\frac{1}{|z|} + 1} + 1}{4\sqrt{\log\frac{1}{|z|}}}\right)^n$$

and both are strongly greater than 1. However, $D(x, 0, f) = D_p(x, 0, f) = 1$ at any point of the punctured unit ball $\mathbb{B}^n(0, 1) \setminus \{0\}$; see again [14]. Observe that the equality of either inner or outer dilatation to 1 in a domain guarantees that the mapping is conformal in this domain.

Following [21], we say that $f : D \to \mathbb{R}^n$ is *finitely Lipschitz* if $L(x, f) < \infty$ for all $x \in D$ and *finitely bi-Lipschitz* if $0 < l(x, f) \leq L(x, f) < \infty$ for all $x \in D$, where

$$L(x, f) := \limsup_{y \to x} \frac{|f(x) - f(y)|}{|x - y|}, \qquad l(x, f) := \liminf_{y \to x} \frac{|f(x) - f(y)|}{|x - y|}.$$

Note that the finite Lipschitzness provides an essentially wider class of mappings than Lipschitz ones. It can be illustrated by simple examples. Moreover even finite bi-Lipschitz mapping can fail to be quasiconformal although every bi-Lipschitz mapping with the Lipschitz norm L is obviously L^2-quasiconformal.

In addition to the above mappings, we consider the classes of mappings of finite metric and area distortions. Firstly, we recall one of essential notions in the mapping theory called the Lusin (N)-property with respect to the corresponding measure. Usually, this condition is also called the absolute continuity in measure.

We recall that a mapping f possesses the Lusin (N)-property with respect to the k-dimensional Hausdorff measure $\mathcal{H}^k$ if $\mathcal{H}^k f(E) = 0$ whenever $\mathcal{H}^k E = 0$. Conversely, f satisfies the Lusin (N^{-1})-property if for any E, $\mathcal{H}^k E = 0$, we have $\mathcal{H}^k f^{-1}(E) = 0$.

We say that a mapping $f : D \to \mathbb{R}^n$ is of *finite metric distortion*, abbr. $f \in$ FMD, if f admits the Lusin (N)-property with respect to the n-dimensional Hausdorff measure and

$$0 < l(x, f) \leq L(x, f) < \infty \qquad \text{almost everywhere (a.e.)},$$

cf. [21].

The Lusin (N)-property with respect the n-dimensional Hausdorff measure cannot guarantee that the same property holds for any smaller integer k, $k = 1, \dots, n-1$. Here we refer to a recent paper [24] where it was shown that a quasiconformal mapping in $\mathbb{R}^3$ does not preserve the Lusin (N^{-1})-property with respect to $\mathcal{H}^2$.

For a Borel function $\rho : \mathbb{R}^n \to [0, \infty]$, its *integral over a k-dimensional surface* $\mathcal{S}$ (a continuous mapping $\mathcal{S} : D_{\mathcal{S}} \to \mathbb{R}^n$, $D_{\mathcal{S}}$ is a domain in $\mathbb{R}^k$, $k = 1, \dots, n-1$) is determined by

$$\int\limits_{\mathcal{S}} \rho \, d\mathcal{A} := \int\limits_{\mathbb{R}^n} \rho(y) \, N(\mathcal{S}, y) \, d\mathcal{H}^k y \,,$$

where $N(\mathcal{S}, y)$ stands for the multiplicity function of $\mathcal{S}$, namely, the multiplicity of covering the point y by the surface $\mathcal{S}$, $N(\mathcal{S}, y) = \operatorname{card} \mathcal{S}^{-1}(y)$, which is lower semicontinuous and, therefore, is measurable with respect to the Hausdorff measure $\mathcal{H}^k$; see [25, p. 160], cf. [21, p. 177].

A Borel function $\rho : \mathbb{R}^n \to [0, \infty]$ is called *admissible* for the family of k-dimensional surfaces Γ in $\mathbb{R}^n$, $k = 1, 2, \dots, n-1$, abbr. $\rho \in \operatorname{adm} \Gamma$, if

$$\int\limits_{S} \rho^k \, d\mathcal{A} \geq 1 \qquad \forall \mathcal{S} \in \Gamma \,.$$

By the *k-dimensional Hausdorff area* of a Borel set B in $\mathbb{R}^n$ (or simply *area* of B in the case $k = n-1$) associated with the surface $\mathcal{S} : \omega \to \mathbb{R}^n$, we mean

$$\mathcal{A}_{\mathcal{S}}(B) = \mathcal{A}^k_{\mathcal{S}}(B) := \int\limits_{B} N(\mathcal{S}, y) \, d\mathcal{H}^k y \,,$$

(cf. [10, Ch. 3.2.1]). The surface $\mathcal{S}$ is called *rectifiable (quadrable)*, if $\mathcal{A}_{\mathcal{S}}(\mathbb{R}^n) < \infty$ (see, e.g., [21, Ch. 9.2]).

The *modulus of family* Γ (conformal modulus) is defined by

$$\mathcal{M}(\Gamma) := \inf_{\rho \in \operatorname{adm} \Gamma} \int\limits_{D} \rho^n(x) \, dm(x) \,. \tag{2.2}$$

Replacing the exponent n in (2.2) by real p, $p \geq 1$, we arrive at the quantity which is called the p-modulus $\mathcal{M}_p(\Gamma)$ of the family Γ.

We say that a property P holds a.e. $\mathcal{S} \in \Gamma$, if the corresponding modulus of a subfamily Γ_* of $\mathcal{S} \in \Gamma$ for which P is not true vanishes.

We also say that a mapping $f : D \to \mathbb{R}^n$ has the *(A_k)-property* if the following two conditions hold:

$(A_k^{(1)})$: for a.e. k-dimensional surface $\mathcal{S}$ in D, the restriction $f|_{\mathcal{S}}$ has the (N)-property with respect to area;

$(A_k^{(2)})$: for a.e. k-dimensional surface $\mathcal{S}'$ in $D' = f(D)$, the restriction $f|_{\mathcal{S}}$ has the (N^{-1})-property for each lifting $\mathcal{S}$ of $\mathcal{S}'$ with respect to area.

Here a surface $\mathcal{S}$ in D is a *lifting* of a surface $\mathcal{S}'$ in D' under a mapping $f : D \to D'$ if $S' = f \circ \mathcal{S}$.

Recall that a mapping $f : D \to \mathbb{R}^n$ is of *finite area distortion in dimension* $k = 1, \ldots, n-1$, abbr. $f \in \mathrm{FAD}_k$, if $f \in$ FMD and has the (A_k)-property; see [19]. We also say that a mapping $f : D \to D'$ is of *finite area distortion*, abbr. $f \in$ FAD, if $f \in \mathrm{FAD}_k$ for every $k = 1, \ldots, n-1$.

To illustrate homeomorphisms of finite metric and area distortions comparing with the quasiconformality and Lipschitzness, we present some examples given in [5].

The stretching

$$f_1(x) = x(1 - \log|x|)$$

is an automorphism of the punctured unit ball $\mathbb{B}^n \setminus \{0\}$. For this homeomorphism, one gets

$$L(x, f_1) = 1 + \log\frac{1}{|x|}, \quad l(x, f_1) = \log\frac{1}{|x|},$$

$$J_{f_1}(x) = \left(1 + \log\frac{1}{|x|}\right)^{n-1} \log\frac{1}{|x|},$$

and f_1 is both an FMD-homeomorphism and FAD-homeomorphism (but not Lipschitz), which has a removable singularity at the origin. However, a direct calculation yields

$$K_O(x, f_1) = \frac{1 + \log\frac{1}{|x|}}{\log\frac{1}{|x|}}, \quad K_I(x, f_1) = \left(\frac{1 + \log\frac{1}{|x|}}{\log\frac{1}{|x|}}\right)^{n-1},$$

which are unbounded in $\mathbb{B}^n \setminus \{0\}$, and, therefore, f_1 fails to be quasiconformal. This mapping is of finite distortion; see, e.g., [18].

The following example provides a classical example of quasiconformal mapping:

$$f_2(x) = |x|^{\alpha-1}x, \qquad 0 < \alpha < 1,$$

which also has a removable singularity at the origin and can be extended to a homeomorphism of the whole unit ball $\mathbb{B}^n$. Indeed, for f_2, we have

$$L(x, f_2) = |x|^{\alpha-1}, \quad l(x, f_2) = \alpha|x|^{\alpha-1}, \quad J_{f_2}(x) = \alpha|x|^{n(\alpha-1)},$$

which imply that f_2 is a diffeomorphism with $0 < l(x, f_2) < L(x, f_2) < \infty$ in $\mathbb{B}^n \setminus \{0\}$. Note also that $K_O(x, f_2) = 1/\alpha$ and $K_I(x, f_2) = 1/\alpha^{n-1}$, which ensure quasiconformality.

Our last example illustrates a mapping defined in the punctured unit ball $\mathbb{B}^n \setminus \{0\}$ by

$$f_3(x) = (x_1 \cos\theta - x_2 \sin\theta, x_2 \cos\theta + x_1 \sin\theta, x_3, \ldots, x_n),$$

with $x = (x_1, \ldots, x_n)$ and $\theta = \log(x_1^2 + x_2^2)$. This mapping preserves the volume.

A straightforward calculation yields

$$L(x, f_3) = 1 + \sqrt{2}, \quad l(x, f_3) = \sqrt{2} - 1, \quad J_{f_3}(x) = 1,$$

see [17]. Thus, f_3 is L-bi-Lipschitz with $L = 1 + \sqrt{2}$, and, therefore, it belongs to FMD, FAD, and QC.

The following statement is crucial in our research and provides an upper integral bound for p-modulus.

Theorem 2.1 *Let $f : D \to \mathbb{R}^n$ be an FAD-homeomorphism and $x_0 \in D$ be an arbitrary point. Assume also that $D_p(x, x_0, f)$ is locally integrable in D. Then for each nonnegative measurable function $\eta(t)$, $t \in (r_1, r_2)$, such that $\int_{r_1}^{r_2} \eta(t)dt = 1$, the following inequality:*

$$\mathcal{M}_p(f(\Gamma)) \leq \int_A \eta^p(|x - x_0|)\, D_p(x, x_0, f)\, dm(x), \tag{2.3}$$

holds for any spherical ring $A = A(x_0, r_1, r_2) = \{x \in \mathbb{R}^n : r_1 < |x - x_0| < r_2\} \subset D$, where Γ stands for the family of curves joining the boundary components of ring A in A.

Proof Let η be a Borel function in $[r_1, r_2]$ satisfying $\int_{r_1}^{r_2} \eta(t)dt = 1$, and define

$$\rho^*(y) = \eta(|x - x_0|) \left(\frac{D_p(x, x_0, f)}{J_f(x)} \right)^{1/p} \circ f^{-1}(y)$$

if f is differentiable and $D_p(x, x_0, f)/J_f(x) \neq 0$ at $x = f^{-1}(y)$ and $\rho^*(y) = \infty$ otherwise at $y \in f(A)$, extending $\rho^*(y) = 0$ outside $f(A)$. This function is admissible for the family $f(\Gamma)$. Next arguing similarly to [14, Theorem 1] and [21, Lemma 10.2], we prove the theorem assertion. □

Note that the inequality (2.3) determines a class of mappings which are called ring Q-homeomorphisms with respect to p-modulus. This class was introduced for the conformal case in [26]; see also [21]. The nonconformal case has been initially considered in [12]. For studying the weighted modulus inequalities, we also refer to [7–9].

The following theorem provides the upper/lower bounds for the Jacobian in accordance to the range of p.

Theorem 2.2 *Let $f : D \to \mathbb{R}^n$ be an FAD-homeomorphism and $x_0 \in D$ be a point. Suppose that $D_p(\cdot, x_0, f) \in L^1_{\text{loc}}(D)$, $p \geq 1$. If $1 \leq p < n$, $J_f(x) \leq K_1 D_p^{\frac{n}{n-p}}(x, x_0, f)$ a.e. For $p > n$, $J_f(x) \geq K_2 D_p^{\frac{n}{n-p}}(x, x_0, f)$. Here K_1, K_2 are positive constants depending on n and p only.*

Proof The proof follows the lines of [11, Lemma 6]; cf. [15, Theorem 4.1]. □

Assuming a special growth estimate for the integral in (2.3) and choosing $\eta(t) = 1/t \log(r_2/r_1)$, one derives the following upper bound for the n-dimensional Hausdorff measure of the image of ball $B(x_0, r)$.

Theorem 2.3 *Let $f : D \to \mathbb{R}^n$ be an FAD-homeomorphism and $x_0 \in D$ be an arbitrary point. Suppose that $D_p(x, x_0, f) \in L^1(D)$ with $1 < p < n$, and that there is $C_{x_0} > 0$, $\kappa \in [0, p)$ such that for any $0 < r_1 < r_2 < d_0 = \text{dist}(x_0, \partial D)$ the following bound:*

$$\int\limits_{A(x_0, r_1, r_2)} \frac{D_p(x, x_0, f)\, dm(x)}{|x - x_0|^p} \leq C_{x_0} \log^{\kappa} \left(\frac{r_2}{r_1}\right),$$

holds. Then

$$\mathcal{H}^n f(B(x_0, r)) \leq C_1 \cdot C_{x_0}^{\frac{n}{n-1}} \log^{\frac{(\kappa - p)n}{n-p}} \left(\frac{1}{r}\right).$$

holds for all $r < \delta_0 \leq \min\{1, \text{dist}^2(x_0, \partial D)\}$, where $C_1 > 0$ depends only on n, p, x_0, and κ.

Proof By Theorem 2.1, we obtain that any FAD-homeomorphism is a ring Q-homeomorphism with respect to p-modulus with $Q(x) = D_p(x, x_0, f)$. Then using Lemma 1 and Theorem 2 from [27] completes the proof. □

3 Main Results

In this chapter we present the main results related to the Hausdorff measures of the images under homeomorphisms of finite area distortion.

Theorem 3.1 *Let $f : D \to \mathbb{R}^n$ be an FAD-homeomorphism satisfying the assumptions of Theorem 2.3. Then for $n - 1 < p < n$,*

$$\text{diam}\, f\left(\overline{B(x_0, r)}\right) \leq C_1 \cdot C_{x_0}^{\frac{1}{n-1}} \log^{\frac{(\kappa - p)}{n-p}} \left(\frac{1}{r}\right)$$

for all $x \in B(x_0, \delta_0)$, $\delta_0 \in (0, \min\{1, \text{dist}^4(x_0, \partial D)\})$.

Proof The proof follows the lines of Theorem 3 from [27]. □

Theorem 3.2 *Let $f : D \to \mathbb{R}^n$ be an FAD-homeomorphism, $E \subset D$ be a Borel set, and $x_0 \in D$ be a point. Suppose that $D_p(\cdot, x_0, f) \in L_{\mathrm{loc}}^{\frac{n}{n-p}}(D)$, $p \geq 1$. If $1 \leq p < n$, then*

$$\mathcal{H}^n f(E) \leq K_1 \int\limits_E D_p^{\frac{n}{n-p}}(x, x_0, f)\, dm(x)\,. \tag{3.1}$$

If $p > n$, then

$$\mathcal{H}^n f(E) \geq K_2 \int\limits_E D_p^{\frac{n}{n-p}}(x, x_0, f)\, dm(x)\,.$$

Proof The required estimates for the n-dimensional Hausdorff measure of the image of E under FAD-homeomorphisms follow from the corresponding bounds for the Jacobian given in Theorem 2.3. □

Theorem 3.3 *Let $f : D \to D'$ be an FAD-homeomorphism and $x_0 \in D$ be a point. Suppose that $D_p(\cdot, x_0, f) \in L_{\mathrm{loc}}^{\frac{n}{n-p}}(D)$, $1 \leq p < n$, and $B(x_0, r)$ is a ball, $\overline{B(x_0, r)} \subset D$. Then for each $\varepsilon > 0$, there exists an open polyhedron $G' \supset f(B(x_0, r))$ such that its each point lies with distance ε of $f(B(x_0, r))$ and*

$$\mathcal{H}^{n-1}\left(\partial G'\right) \leq K(1+\varepsilon)\left(\fint_A D_p^{\frac{n}{n-p}}(x, x_0, f)\right)^{\frac{p(n-1)}{n}} \mathcal{H}^{n-1}(\partial U)\,, \tag{3.2}$$

where K depends on n and p only and $A = A(x_0, r, R)$ is a spherical ring, $\overline{A} \subset D$.

Proof The proof of Theorem 3.3 involves some techniques and estimates. First we apply the p-capacities of ring condensers $f(A)$ and their relations with the p-moduli of curves families which join the boundary components of $f(A)$ in $f(A)$. The next step relates to estimating the p-capacities by the $(n-1)$-Hausdorff measure of piecewise smooth surfaces which contain $f(A)$ and the n-dimensional Hausdorff measures of $f(A)$. Finally, the estimate (3.1) yields the desired bound (3.2) in the terms of integral averages of the angular dilatation to the power $n/(n-p)$ over the spherical ring $A = A(x_0, r, R)$. □

References

1. E. Afanas'eva, On the boundary behavior of one class of mappings in metric spaces. Ukrainian Math. J. **66**(1), 16–29 (2014)
2. E. Afanas'eva, Generalized quasi-isometries on smooth Riemannian manifolds. Math. Notes **102**(1–2), 12–21 (2017)

3. E. Afanas'eva and A. Golberg, Finitely bi-Lipschitz homeomorphisms between Finsler manifolds. Anal. Math. Phys. **10**(4), Paper No. 48, 1–16 (2020)
4. E. Afanas'eva, A. Golberg, Topological mappings of finite area distortion. Anal. Math. Phys. **12**(2), Paper No. 54, 1–29 (2022)
5. E. Afanas'eva, A. Golberg, Homeomorphisms of finite metric distortion between Riemannian manifolds. Comput. Methods Funct. Theory **22**(4), 755–780 (2022)
6. K.Y. Chen, A.N. Fang, Bi-Lipschitz maps in Q-regular Loewner spaces. Acta Math. Sin. (Engl. Ser.) **24**(9), 1555–1568 (2008)
7. M. Cristea, Eliminability results for mappings satisfying generalized modular inequalities. Complex Var. Elliptic Equ. **64**(4), 676–684 (2019)
8. M. Cristea, On Poleckii-type modular inequalities. Complex Var. Elliptic Equ. **66**(11), 1818–1838 (2021)
9. M. Cristea, Cluster sets theorems on metric measure spaces. Anal. Math. Phys. **12**(6), Paper No. 139, 1–19 (2022)
10. H. Federer, *Geometric Measure Theory*. Die Grundlehren der mathematischen Wissenschaften, Band 153. (Springer, New York, 1969) (2nd edn. 1996)
11. F.W. Gehring, Lipschitz mappings and the p-capacity of rings in n-space, in *Advances in the Theory of Riemann Surfaces* (Proc. Conf., Stony Brook, 1969). Annals of Mathematics Studies, vol. 66 (Princeton University Press, Princeton, 1971), pp. 175–193
12. A. Golberg, Differential properties of (α, Q)-homeomorphisms, in *Further Progress in Analysis* (World Scientific Publishing, Hackensack, 2009), pp. 218–228
13. A. Golberg, Directional dilatations in space. Complex Var. Elliptic Equ. **55**(1–3), 13–29 (2010)
14. A. Golberg, R. Salimov, Homeomorphisms Lipschizian in the mean, in *Complex Analysis and Potential Theory with Applications* (Cambridge Science Publishing, Cambridge, 2014), pp. 95–111
15. A. Golberg, R. Salimov, E. Sevost'yanov, Estimates for Jacobian and dilatation coefficients of open discrete mappings with controlled p-module. Complex Anal. Oper. Theory **11**, 1521–1542 (2017)
16. V. Gol'dshtein, A. Ukhlov, About homeomorphisms that induce composition operators on Sobolev spaces. Complex Var. Elliptic Equ. **55**(8–10), 833–845 (2010)
17. V.Y. Gutlyanskiĭ, A. Golberg, On Lipschitz continuity of quasiconformal mappings in space. J. Anal. Math. **109**, 233–251 (2009)
18. S. Hencl, P. Koskela, Lectures on mappings of finite distortion, in *Lecture Notes in Mathematics*, vol. 2096 (Springer, Cham, 2014)
19. D. Kovtonyuk, V. Ryazanov, On the theory of mappings with finite area distortion. J. Anal. Math. **104**, 291–306 (2008)
20. V. Kud'yavin, A. Golberg, Mean coefficients of quasiconformality of a pair of domains. Ukrainian Math. J. **43**(12), 1594–1597.
21. O. Martio, V. Ryazanov, U. Srebro, E. Yakubov, *Moduli in Modern Mapping Theory*. Springer Monographs in Mathematics (Springer, New York, 2009)
22. P. Mattila, *Rectifiability—A Survey*. London Mathematical Society Lecture Note Series, vol. 483 (Cambridge University Press, Cambridge, 2023)
23. P. Mattila, M. Vuorinen, Linear approximation property, Minkowski dimension, and quasiconformal spheres. J. Lond. Math. Soc. (2) **42**(2), 249–266 (1990)
24. D. Ntalampekos, M. Romney, On the inverse absolute continuity of quasiconformal mappings on hypersurfaces. Am. J. Math. **143**, 1633–1659 (2021)
25. T. Rado, P.V. Reichelderfer, Continuous transformations in analysis, with an introduction to algebraic topology. Die Grundlehren der mathematischen Wissenschaften in Einzeldarstellungen mit besonderer Berücksichtigung der Anwendungsgebiete, Bd. LXXV (Springer, Berlin/Göttingen/Heidelberg, 1955)
26. V. Ryazanov, U. Srebro, E. Yakubov, On ring solutions of Beltrami equations. J. Anal. Math. **96**, 117–150 (2005)
27. R.R. Salimov, On ring Q-mappings with respect to a nonconformal modules. (Russian) Far Eastern Math. J. **14**(2), 257–269 (2014)

28. R. Salimov, E. Sevost'yanov, A. Ukhlov, Capacity inequalities and Lipschitz continuity of mappings, arXiv preprint arXiv:2302.13302 (2023)
29. J. Väisälä, *Lectures on n-Dimensional Quasiconformal Mappings*. Lecture Notes in Mathematics, vol. 229 (Springer, Berlin/New York, 1971)

The Principal Symbol of Elliptic Differential Operators on Lie Groups and Homogeneous Spaces

Alexander Cardona and Santiago Gómez Cobos

Abstract We associate a principal symbol with elliptic differential operators acting on sections of vector fibrations over Lie groups and homogeneous spaces. This principal symbol is an element in the ring of virtual representations of the group, and, in particular instances, we show that it gives insights on the computation of spectral invariants linked to this type of operators.

2010 Mathematics Subject Classification 47G30, 14M17, 43A65

1 Introduction

Consider an elliptic pseudodifferential operator $P : \Gamma(E) \to \Gamma(F)$ acting on sections of finite-rank vector bundles over a manifold M, $E \to M$ and $F \to M$. It is well-known that, in the elliptic case, the principal symbol of the operator defines an invertible element in $\Gamma(T^*M, \mathrm{Hom}(\pi^*E, \pi^*F))$ from which the index of P can be computed , where $\pi^*E \to M$ and $\pi^*F \to M$ denote the corresponding pullback bundles with respect to the cotangent bundle projection $\pi : T^*M \to M$. Actually, the principal part of the symbol defines an element in $K(M)$, the main ingredient for the local index formula of Atiyah and Singer [1]. If $M = G$, a compact connected (and simply connected) Lie group, following [6], we can associate with a linear operator $P : C^\infty(G) \to C^\infty(G)$ matrix symbols $\sigma(P) : G \times \widehat{G} \to \bigcup_{\xi \in \widehat{G}} \mathrm{End}(\mathcal{H}_\xi)$

A. Cardona
Departamento de Matemáticas, Universidad de los Andes, Bogotá, Colombia
e-mail: acardona@uniandes.edu.co

S. Gómez Cobos (✉)
Department of Mathematics: Analysis, Logic and Discrete Mathematics, Ghent University, Ghent, Belgium
e-mail: davidsantiago.gomezcobos@ugent.be

M. Ruzhansky, B. Torebek (eds.), *Extended Abstracts MWCAPDE 2023*,
Research Perspectives Ghent Analysis and PDE Center 1,
https://doi.org/10.1007/978-3-031-41665-1_2

by the formula

$$\sigma(P)(x,\xi) = \xi^*(x)(P\xi)(x) \in \mathrm{End}(\mathcal{H}_\xi) \tag{1.1}$$

where $x \in G$ and $\xi \in \widehat{G}$ denotes an irreducible unitary representation $\xi : G \to \mathrm{End}(\mathcal{H}_\xi)$. In this note we will consider differential operators

$$D : \Gamma(E) \to \Gamma(F)$$

acting on sections of vector fibrations over the Lie group G, on the one hand, and over homogeneous spaces G/H, on the other hand, and we will define the principal symbols of such operators as elements of $R(G)$, the virtual representation ring of the Lie group G (i.e. the Grothendieck group associated with the semigroup $\widehat{G}$). This constructions give a first insight on the more general question of computing spectral (index-like) invariants for elliptic differential operators of this type which, to our knowledge, is well-known only in particular cases from the point of view of the global pseudodifferential calculus of Ruzhansky and Turunen [6]. We will present particular instances in which the elements in the representation ring contain important information we can use to compute spectral invariants, namely, the case of differential forms on Lie groups and on spheres, seen as homogeneous spaces; more complete treatment of this problem will be presented elsewhere.

2 Vector Bundles over Lie Groups Defined by Representations

In this sections we will consider a compact, connected and simply connected Lie group G as an homogeneous space, following [3]; this will allow us to associate a "principal symbol" $\sigma(D)$ with an elliptic differential operator $D : \Gamma(E) \to \Gamma(F)$ for vector bundles associated with representations.

Consider the Lie group $\mathbb{G} = G \times G$ with its natural structure and the diagonal $\mathrm{diag}(G) = \{(g,g) : g \in G\} \subset \mathbb{G}$, for which clearly $\mathrm{diag}(G) \cong G$. Let us take the following smooth map:

$$\lambda : \mathbb{G} \to G$$

$$(x,y) \mapsto xy^{-1}$$

which induces the exact sequence

$$\{e\} \to \mathrm{diag}(G) \to \mathbb{G} \xrightarrow{\lambda} G \to \{e\}$$

so that, as manifolds, $G \cong \mathbb{G}/\mathrm{diag}(G)$. This realization of G as a quotient of the double $\mathbb{G}$ by the diagonal gives a natural way to associate, to any finite-dimensional representation of G,

$$\xi : G \to \mathrm{End}(\mathcal{H}_\xi)$$

a vector bundle $E_\xi \xrightarrow{\pi_\xi} G$ over the Lie group G as follows. Consider the trivial bundle

$$\mathbb{G} \times \mathcal{H}_\xi \to \mathbb{G} \tag{2.1}$$

and define on its total space the action of $\mathrm{diag}(G)$ given by

$$\bar{\xi} : \mathrm{diag}(G) \times (\mathbb{G} \times \mathcal{H}_\xi) \to \mathbb{G} \times \mathcal{H}_\xi$$
$$((g, g), (x, y, v)) \mapsto (xg^{-1}, yg^{-1}, \xi(g)v)$$

using ξ and the natural identification $\mathrm{diag}(G) \cong G$. Then, taking the quotient of (2.1) by this action, we obtain a vector bundle denoted by

$$E_\xi := \mathbb{G} \times_\xi \mathcal{H}_\xi \to G \cong \mathbb{G}/\,\mathrm{diag}(G).$$

Particular instances of this construction give rise to some important vector bundles over Lie groups G, for example, using the adjoint action of G over its Lie algebra $\mathfrak{g}$, $\mathrm{Ad} : G \to \mathrm{End}(\mathfrak{g})$, it follows that

$$TG \cong \mathbb{G} \times_{\mathrm{Ad}} \mathfrak{g}$$

and, similarly, using the co-adjoint action $\mathrm{Ad}^* : G \to \mathrm{End}(\mathfrak{g}^*)$, $T^*G \cong \mathbb{G} \times_{\mathrm{Ad}^*} \mathfrak{g}^*$ and, in general, $\bigwedge^p T^*G \cong \mathbb{G} \times_{\wedge^p \mathrm{Ad}^*} \bigwedge^p \mathfrak{g}^*$.

Notice that these bundles are trivial, so that the corresponding spaces of (square-integrable) sections have a very simple structure, namely, $\mathfrak{X}(G) \cong L^2(G) \otimes \mathfrak{g}$, $\Gamma(T^*G) \cong L^2(G) \otimes \mathfrak{g}^*$, and $\Omega^p(G) \cong L^2(G) \otimes \bigwedge^p \mathfrak{g}^*$, respectively, which allows us to use Peter-Weyl theorem to find convenient decompositions of them in terms of irreducible representations of G.

3 Elliptic Differential Operators and Principal Symbols

Consider an element $[\xi] - [\eta] \in R(G)$, i.e. a virtual representation of G. Then, by the construction before, we can naturally associate an element in $K(G)$, the (virtual) vector bundle $[E_\xi] - [E_\eta]$. In this situation, there is a natural projection

$$\pi : \mathbb{G} \times_{\mathrm{Ad}^*} \mathfrak{g}^* \to G$$
$$[(x, y), \alpha] \mapsto xy^{-1}$$

which allows us to induce bundles

$$\begin{array}{ccc} \pi^* E_\xi & & \pi^* E_\eta \\ & \searrow \quad \swarrow & \\ & T^*G & \end{array}$$

Consider now the principal symbol of a differential operator

$$D : \Gamma(E_\xi) \to \Gamma(E_\eta),$$

i.e. a section $\sigma(D) \in \Gamma(T^*G, \mathrm{Hom}(\pi^* E_\xi, \pi^*_\eta))$. Since $T^*G \cong \mathbb{G} \times_{\mathrm{Ad}^*} \mathfrak{g}^*$ is trivial $\sigma(D) \in C^\infty(G) \otimes \mathrm{Hom}(E_\xi \otimes \mathfrak{g}^*, E_\eta \otimes \mathfrak{g}^*)$, and D is *elliptic* if (away the zero section) $\sigma(D)$ is invertible.

Definition 3.1 The principal symbol of D is the virtual representation $[\xi] - [\eta] \in R(G)$. Moreover, since there is a one-to-one correspondence between irreducible unitary representations and their highest weights, which are some special elements in the dual Lie algebra of its maximal torus, we can write the symbol as $\mu_\xi - \mu_\eta \in \mathfrak{t}^*$.

Remark 3.2 Notice that the element in $K(G)$ associated with D is $[E_\xi \otimes \mathrm{Ad}^*] - [E_\eta \otimes \mathrm{Ad}^*]$, but the co-adjoint representations factors out as constant coefficient for the principal symbol.

Example 3.3 Consider the de Rham operator on differential forms on G, namely,

$$d : \Omega^*(G) \to \Omega^{*+1}(G).$$

Since $\Omega^p(G) \cong C^\infty(G) \otimes \bigwedge^p \mathfrak{g}^*$, using the Peter-Weyl theorem, we can decompose p-forms on G as $\Omega^p(G) \cong \sum_\xi \mathcal{H}_\xi \otimes \mathcal{H}_\xi^* \otimes \bigwedge^p \mathfrak{g}^*$, giving an identification of matrix coefficients of the representation $\wedge^p \mathrm{Ad}^*$ with p-forms. The principal symbol of $d : \Omega^p(G) \to \Omega^{p+1}(G)$ is the element in $R(G)$ given by $[\wedge^p \mathrm{Ad}^*] - [\wedge^{p+1} \mathrm{Ad}^*]$, which corresponds geometrically to "$\mathrm{Ad}^* \wedge$" (exterior multiplication by the co-adjoint representation). These representations can be parametrized using highest weights, which becomes useful when it comes to compute invariants from the principal symbols in $R(G)$.

4 Vector Bundles over Homogeneous Spaces and Bott's Induction

In this section we present the construction of principal symbols for homogeneous operators on homogeneous spaces following Bott [2] and illustrate its use in some examples. Let G be a compact Lie group, and H be a closed subgroup of G, $\rho_i : H \to \mathrm{GL}(V_i)$ representations of H for $i = 1, 2$.

Definition 4.1 We say that a differential operator $D : \Gamma(E) \to \Gamma(F)$ acting on sections of vector bundles is homogeneous if it satisfies the following conditions:

(1) The manifold M is an homogeneous space, i.e. $M \cong G/H$.
(2) E and F are induced by the representations ρ_i, respectively. This means that $E = G \times_H V_{\rho_1}$ and $F = G \times_H V_{\rho_2}$.
(3) D commutes with the induced action of G in $\Gamma(E)$ and $\Gamma(F)$, i.e. $g.(Ds) = D(g.s)$.

Remark 4.2 Note that in this situation G acts naturally on $\operatorname{Ker} D$ and $\operatorname{Coker} D$ since they are subspaces of sections, so they are G-modules, i.e. representations of G. Hence, they define elements on the representation ring $R(G)$, and the index of D is coming from the "principal symbol"

$$\chi(D) = [\operatorname{Ker} D] - [\operatorname{Coker} D] \in R(G)$$

since

$$\operatorname{ind}(D) = \dim \chi(D).$$

In this setting Bott [2] proved an index theorem, in which he obtained that due to the symmetry of the problem, the information related to the index is concentrated on what he called the homogeneous symbol. Such a result is the following:

Theorem 4.3 (Bott [2]) *Let D be a homogeneous elliptic differential operator $D : \Gamma(E) \to \Gamma(F)$ where the bundles are induced by representations of H ρ_1 and ρ_2, respectively. Define the homogeneous symbol $\sigma(D) := [\rho_1] - [\rho_2] \in R(H)$. If $\iota_* : R(H) \to \hat{R}(G)$ is formal induction, then*

(1) $\iota_*\sigma(D)$ is a finite element of $\hat{R}(G)$.
(2) $\chi(D) = \iota_*\sigma(D)$.

The key part of this theorem resides in the use of formal induction, which is a generalization of classical induction that appears, for example, in the Frobenius reciprocity theorem. We believe this process could be implemented from the Harmonic analysis point of view as well, for this reason we will explain it in detail.

Let $R(G)$ be the representation ring of G, i.e. virtual differences of isomorphism classes of finite dimensional representations of G. Since G is compact, all the representations are reducible so that if $\{W_\alpha\}$ is a complete set of irreducible nonequivalent representations, then the induced elements $\{w_\alpha = [W_\alpha]\}$ are a free basis of $R(G)$. The occurrences of one representation into other, so-called intertwining number,

$$\tau(A, B) = \dim \operatorname{Hom}_G(A, B)$$

defines a symmetric bi-linear form on $R(G)$, which we denote by $\langle \cdot, \cdot \rangle_G$ and our canonical basis is orthonormal with respect to it. We define the formal group $\hat{R}(G)$ as the possibly infinite linear combinations $\sum a_\alpha w_\alpha, a_\alpha \in \mathbb{Z}$ of the canonical basis. So we can see $R(G)$ included in $\hat{R}(G)$ as the finite elements. Given a homomorphism $\varphi : H \to G$ (not only the inclusion), we define the formal induced representation $\varphi_* : R(H) \to \hat{R}(G)$ as the extension to $R(H)$ of the function

$$\rho \mapsto \sum \langle \varphi^* w_\alpha, [\rho] \rangle_H w_\alpha \text{ for } \rho \text{ a representation of } H.$$

This basically says that you construct a new "representation" by weighing correctly the information carried by φ.

Once the formal induction is understood, we proceed to write the intentions of this project which, in a few words, seeks to adapt this method to the context of Ruzhansky and Turunen's quantization. We can formulate the global pseudodifferential theory in [6] in the generality of Definition 4.1 as follows. We claim that one can associate a global symbol with a homogeneous differential operator $D : \Gamma(E) \to \Gamma(F)$ using again Fourier analysis (in this case vector-valued) since the space of sections of such bundles can be identified with the following space of invariant functions:

$$\Gamma(E) = \left\{ V_\rho\text{-valued } L^2 \text{ functions on } G \,\middle|\, \begin{array}{l} f(gh) = \rho(h)^{-1} f(g) \text{ for all} \\ h \in H \text{ and almost every } g \in G \end{array} \right\}.$$

Therefore, there would exists a global symbol $\sigma_D(x, \xi) \in \mathrm{End}(\mathcal{H}_\xi) \otimes \mathrm{Hom}(E_x, F_x)$, where $x \in G$ and $\xi \in \widehat{G}$, such that the operator D can be rewritten as

$$Df(x) = \sum_{\xi \in \widehat{G}} \dim(\xi) \operatorname{Tr}\left(\xi(x) \sigma_D(x, \xi) \widehat{f}(\xi)\right), \text{ for } f \in \Gamma(E).$$

So a natural question in this point is what would be the principal part of the global symbol σ_D. Due to the big amount of information contained in σ_D and the lack of algebraic structure on the phase space $G \times \widehat{G}$, there is not an obvious answer to this problem. We think that Bott index theorem is giving us hints of how to deal with this problem, so in the next section we present some examples that shows how the theory works and them will play the role of toy models to verify future results on the search of a principal part of global symbols.

4.1 Laplacian on G

First of all, let's take a look to the simplest case of homogeneous elliptic differential operators. Suppose $H = \{e\}$ is the trivial subgroup; therefore, we have the following picture:

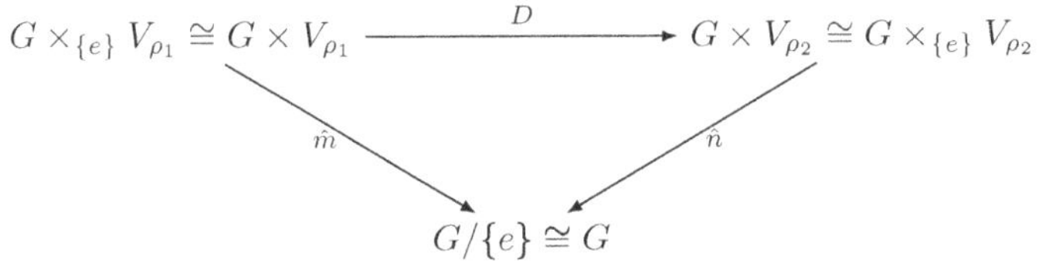

This means that the induced operator on sections is a classical left invariant elliptic differential operator acting on vector valued functions $D : C^\infty(G, V_{\rho_1}) \to C^\infty(G, V_{\rho_2})$. Moreover, if we ask ρ_1 and ρ_2 to be one dimensional, we will have

$$D : C^\infty(G) \to C^\infty(G) \tag{4.1}$$

and according to Theorem 4.3 $\sigma(D) \equiv 0$ since these representations must be the same (representations of trivial subgroup). This immediately implies that the index of this type of operators is always equal to 0, so the question is how to obtain this number from the global symbol defined by Ruzhansky and Turunen.

To ground the ideas a bit, let's see in a little more detail the prototype operator, the Laplacian. Suppose our compact Lie group G is equipped with a bi-invariant Riemannian metric; thus $\mathcal{L}_G$ is the Laplace-Beltrami operator which is a second-order negative definite bi-invariant elliptic differential operator. This operator plays a huge role in the Fourier analysis on the group since elements of $\widehat{G}$ are eigenfunctions of $\mathcal{L}_G$; thus its global symbol is given by [6]

$$\sigma_{\mathcal{L}_G}(x, \xi) = -\lambda_\xi I_\xi$$

where λ_ξ is the eigenvalue of $-\mathcal{L}_G$ and I_ξ is the identity matrix on a Hilbert space of dimension $\dim \xi$. On the other hand, let us compute in detail the Bott induction process for a representation $\rho : \{e\} \to \mathrm{GL}_n(V_\rho)$, which is nothing but a constant map to I_n.

$$\iota_* \rho = \sum \langle \iota^* w_\alpha, \rho \rangle_{\{e\}} w_\alpha = \sum \langle w_\alpha(e), \rho \rangle_{\{e\}} w_\alpha = w_\alpha^n,$$

where w_α^n means the unitary irreducible representation of G of dimension n. Then, we would like to relate $\iota_* \sigma(\mathcal{L}_G) = w_\alpha^1 - w_\alpha^1 \equiv 0 \in R(G)$ with $\sigma_{\mathcal{L}_G}(x, \xi) = -\lambda_\xi I_\xi \in \mathrm{End}(\mathcal{H}_\xi)$.

4.2 *De Rham Operator on* $\mathbb{S}^n$

Recall that the Laplace-Beltrami operator $\Delta : \Omega^*(M) \to \Omega^*(M)$ can be defined on the graded algebra of forms using the exterior derivative and its adjoint; this is $\Delta = dd^* + d^*d$. On the other hand, this definition is equivalent to $\Delta = (d + d^*)^2$ since $d^2 = 0$ and $(d^*)^2 = 0$, which implies that $D = d + d^* : \Omega^*(M) \to \Omega^*(M)$ is the Dirac operator associated with this Laplacian (the "square root" of the Laplacian, also called the Dirac-de Rham operator). This operator D is self-adjoint as well, so its index is zero and doesn't bring too many insights to us. However, we can define a different graduation on this algebra and obtain the geometric meaningful part of this operator. We can separate it in even and odd forms, so

$$\Omega^*(M) = \bigoplus_{k \text{ even}} \Omega^k(M) \oplus \bigoplus_{k \text{ odd}} \Omega^k(M).$$

Henceforth we shall denote $\bigoplus_{k \text{ even}} \Omega^k(M)$ as $\Omega_0(M)$ and $\bigoplus_{k \text{ odd}} \Omega^k(M)$ as $\Omega_1(M)$. Therefore

$$D : \Omega_0(M) \oplus \Omega_1(M) \to \Omega_0(M) \oplus \Omega_1(M),$$

which leads us to think of D as made of two different operators, D restricted to even forms and D restricted to odd forms. We will denote these operators by

$$D^+ : \Omega_0(M) \to \Omega_1(M) \text{ and } D^- : \Omega_1(M) \to \Omega_0(M),$$

and it is well-known that $\operatorname{ind}(D^+) = \operatorname{ind}(D^-) = \chi(M)$, i.e. this operator knows the Euler characteristic of M. For more information about this construction, see [5].

Now we focus on the case $M = \mathbb{S}^n$; we will see how D^+ is an homogeneous elliptic differential operator. Let $G = \mathrm{SO}(n)$, $H = \mathrm{SO}(n-1)$, $V_{\sigma_k} = \bigwedge^k(\mathbb{R}^{n-1})^* \otimes \mathbb{C}$ where $\sigma_k(T)\omega(x) = \omega(T^{-1}x)$. Thus $G/H \cong \mathbb{S}^{n-1}$ and $G \times_H V_{\sigma_k} \cong \bigwedge^k T^*\mathbb{S}^{n-1} \otimes \mathbb{C}$; then sections of this vector bundle are naturally identified with square-integrable k-forms, which we denote Γ_k. Therefore, we can write the exterior derivative naturally in this homogeneous picture

$$\begin{array}{ccc} G \times_H M \cong \bigwedge^k(\mathbb{S}^{n-1})^* \otimes \mathbb{C} & \xrightarrow{\quad d \quad} & \bigwedge^{k+1}(\mathbb{S}^{n-1})^* \otimes \mathbb{C} \cong G \times_H N \\ & \searrow^{\hat{m}} \quad \swarrow^{\hat{n}} & \\ & G/H \cong \mathbb{S}^{n-1} & \end{array}$$

where the induced operator in sections is precisely the exterior derivative $d : \Gamma_k \to \Gamma_{k+1}$.

We can write the $\mathbb{Z}_2$-grading of Ω^* using representations as well. Since Bott's induction process commutes with direct sums, we have

$$\Omega_0(M) = \bigoplus_{k \text{ even}} G \times_{\sigma_k} M, \qquad \Omega_1(M) = \bigoplus_{k \text{ odd}} G \times_{\sigma_k} M,$$

therefore D^+ is an homogeneous elliptic differential operator on vector bundles

$$\begin{array}{ccc} \Omega_0(M) = \bigoplus_{k \text{ even}} G \times_{\sigma_k} M & \xrightarrow{\quad D^+ \quad} & \bigoplus_{k \text{ odd}} G \times_{\sigma_k} M = \Omega_1(M) \\ & \searrow^{\hat{m}} \quad \swarrow^{\hat{n}} & \\ & G/H \cong \mathbb{S}^{n-1} & \end{array}$$

We are in the position to apply Theorem 4.3 to compute the index of D^+:

$$\begin{aligned}
\iota_*\sigma(D^+) &= \sum_\alpha \langle \iota^* w_\alpha, \oplus\sigma_k - \oplus\sigma_{k+1}\rangle_{\mathrm{SO}(n-1)} w_\alpha \\
&= \sum_\alpha \langle w_\alpha|_{\mathrm{SO}(n-1)}, \oplus\sigma_k\rangle_{\mathrm{SO}(n-1)} w_\alpha \\
&\quad - \sum_\alpha \langle w_\alpha|_{\mathrm{SO}(n-1)}, \oplus\sigma_{k+1}\rangle_{\mathrm{SO}(n-1)} w_\alpha \\
&= \sum_{k \text{ even}} \sum_\alpha \langle w_\alpha|_{\mathrm{SO}(n-1)}, \sigma_k\rangle_{\mathrm{SO}(n-1)} w_\alpha \\
&\quad - \sum_{k \text{ odd}} \sum_\alpha \langle w_\alpha|_{\mathrm{SO}(n-1)}, \sigma_k\rangle_{\mathrm{SO}(n-1)} w_\alpha \\
&= \sum_{k \text{ even}} \sum_\kappa \langle \rho_n(\kappa), \pi_k\rangle_{\mathrm{SO}(n)} \rho_n(\kappa) \\
&\quad - \sum_{k \text{ odd}} \sum_\kappa \langle \rho_n(\kappa), \pi_k\rangle_{\mathrm{SO}(n)} \rho_n(\kappa),
\end{aligned}$$

where, following Folland's notation [4], $\rho_n(\kappa)$ denotes the highest-weight representation of $\mathrm{SO}(n)$ corresponding to the weight κ and $\pi_k = \operatorname{ind}\sigma_k$. Theorem A in [4] tells us precisely what numbers $\langle \rho_n(\kappa), \pi_k\rangle_{\mathrm{SO}(n)}$ are nonzero, so using this result we can proceed with the infinite cancellation of terms. We have to consider two cases: $n = 2\nu$ and $n = 2\nu + 1$, where $\nu = \lfloor \frac{n}{2} \rfloor$.

Case $\mathbf{n} = \mathbf{2\nu}$: In this case the cancellation is explained in the diagram

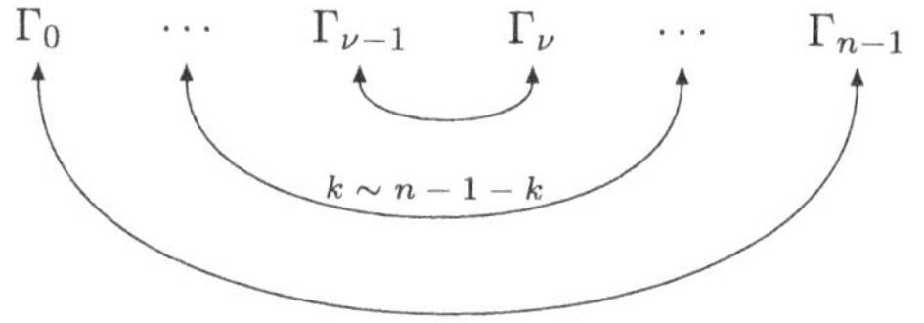

i.e. the representations occurring in Γ_k cancels out with the ones occurring in Γ_{n-1-k}, which is perfect since n is even. Thus $\iota_*\sigma(D^+) = 0$ and $\operatorname{ind}(D^+) = 0$.

Case $\mathbf{n} = \mathbf{2\nu + 1}$: In this case we have to be more careful because at first sight there is not a space cancelling out the representations occurring in Γ_ν. However, if we take a closer look to the decomposition of Γ_ν, we find out that this representations have weight 2, so that both $\Gamma_{\nu-1}$ and $\Gamma_{\nu+1}$ will cancel them. But some representations will remain; thus they will be cancelled by $\Gamma_{\nu-2}$ and $\Gamma_{\nu+2}$ and so on (see the picture below).

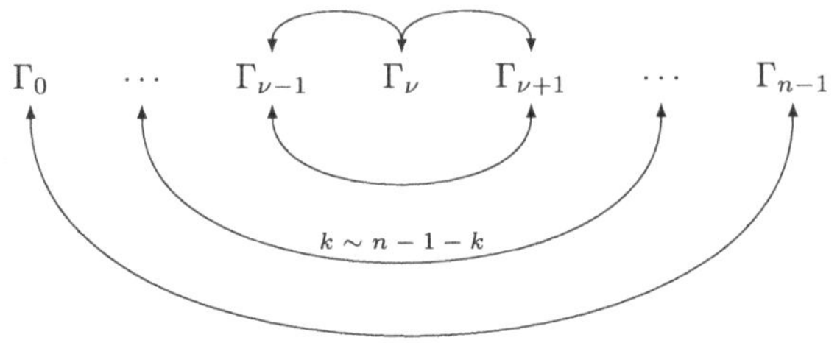

At the end of this process, we will find that the representations coming from the end points Γ_0 and Γ_{n-1} that cannot be cancelled are two times $\rho_n(0_\nu)$, so that $\iota_*\sigma(D^+) = 2\rho_n(0_\nu)$ and $\text{ind}(D^+) = 2$. Thus, we would like to link $\iota_*\sigma(D^+) = 2\rho_n(0_\nu) \in R(\text{SO}(n))$ with $\sigma_{D^+}(x, \xi) \in \text{End}(\mathcal{H}_\xi) \otimes \text{Hom}(V_{\pi_k,x}, V_{\pi_{k+1},x})$. With this computation we obtain the already known Euler characteristic of spheres $\mathbb{S}^n$ (0 or 2, depending on its parity). This calculation is a particular instance of the Dirac-de Rham operator index made by Bott.

Acknowledgments AC was supported by the Faculty of Sciences of the Universidad de los Andes. SGC was supported by the FWO Odysseus 1 grant G.0H94.18N: Analysis and Partial Differential Equations, the Methusalem programme of the Ghent University Special Research Fund (BOF) (Grant number 01M01021).

References

1. M.F. Atiyah, I.M. Singer, The index of elliptic operators. III. Ann. Math. (2) **87**, 546–604 (1968)
2. R. Bott, The index theorem for homogeneous differential operators, in *Differential and Combinatorial Topology*. A Symposium in Honor of Marston Morse (Princeton University Press, Princeton, 1965), pp. 167–186
3. H.D. Fegan, The spectrum of the Laplacian on forms over a Lie group. Pac. J. Math. **90**(2), 373–387 (1980)
4. G.B. Folland, Harmonic analysis of the de Rham complex on the sphere. J. Reine Angew. Math. **398**, 130–143 (1989)
5. A. Mukherjee, *Atiyah-Singer Index Theorem. An Introduction*. Texts and Readings in Mathematics, vol. 69. (Hindustan Book Agency, New Delhi, 2013), xii+267pp
6. M. Ruzhansky, V. Turunen, *Pseudo-Differential Operators and Symmetries: Background Analysis and Advanced Topics* (Birkhäuser-Verlag, Basel, 2010)

On Fractional Quaternionic d-Bar Derivatives

Arran Fernandez, Cihan Güder, and Walaa Yasin

Abstract We introduce a fractional version of the d-bar derivative operator acting on quaternionic functions. Some natural properties of this new operator are shown, including composition properties and its action on certain functions such as Fueter polynomials, which help to demonstrate its suitability compared with other attempts to define fractional quaternionic and Clifford derivatives.

2020 Mathematics Subject Classification 47S05, 26A33

1 Introduction

The study of Clifford algebras brings together concepts from algebra, geometry, and analysis, to create an interdisciplinary field that also has applications in areas ranging from physics to image processing [3, 7]. Within this field, geometric concepts like rotation and reflection can be understood by purely algebraic operations, while algebraic concepts like multiplication can be given a geometric interpretation. Analysis can also be performed in the setting of Clifford algebras, giving rise to Clifford analysis [5] which can be seen as a higher-dimensional extension of complex analysis.

Fractional calculus is a field of study that concerns the extension of the concepts of differentiation and integration to orders beyond $\mathbb{Z}$. Such extensions can be defined in multiple ways, creating a rich variety of theories [9], and applications of fractional models can be found in many fields of science [10]. The essential idea of fractional calculus dates back to 1695, but the field has undergone rapid expansion in the late twentieth and early twenty-first centuries.

A. Fernandez (✉) · C. Güder · W. Yasin
Department of Mathematics, Eastern Mediterranean University, Famagusta, Northern Cyprus, Turkey
e-mail: arran.fernandez@emu.edu.tr; cihan.guder@emu.edu.tr; walaa.yasin@emu.edu.tr

M. Ruzhansky, B. Torebek (eds.), *Extended Abstracts MWCAPDE 2023*,
Research Perspectives Ghent Analysis and PDE Center 1,
https://doi.org/10.1007/978-3-031-41665-1_3

Complex analysis and fractional calculus have been considered in the same breath since the nineteenth century [8], in the sense of using complex analysis to define fractional operators, and introducing complex variables into fractional calculus is still commonly done [9]. On the other hand, little has been done to introduce fractional orders into complex analysis, fractionalising complex analysis rather than complexifying fractional calculus. One of the contributions in this direction was due to the first author, who defined fractional versions of the complex d-bar derivative and Wirtinger derivatives [1].

A particularly important and fundamental Clifford algebra, often considered as the next simplest case after $\mathbb{R}$ and $\mathbb{C}$, is the set $\mathbb{H}$ of quaternions. According to a theorem of Frobenius [4], the only finite-dimensional associative division algebras over $\mathbb{R}$ are precisely $\mathbb{R}$, $\mathbb{C}$, and $\mathbb{H}$, which means these are the only Clifford algebras that are also division algebras, and $\mathbb{H}$ is the only non-commutative one.

A natural question, then, following the development of a fractional complex d-bar derivative [1], is whether this work can be extended to fractional d-bar derivatives in higher-dimensional Clifford algebras or at least in the space of quaternions. Quaternionic analysis has been developed partially in analogy with complex analysis [2], but the only attempts so far to build a fractional version of quaternionic or Clifford analysis [6] have not taken account of the key desired properties of fractional calculus.

Here, we present a short announcement of new research concerning fractional quaternionic analysis, which will be presented more fully in a forthcoming paper.

2 The Main Definition

Before starting to define the fractional quaternionic derivative, first we will show how to find the nth power of the usual d-bar derivative for quaternionic functions. For a quaternion-valued function f of a quaternionic variable $x = x_0 + e_1x_1 + e_2x_2 + e_3x_3$, we have

$$\overline{\partial} f = \frac{\partial f}{\partial x_0} + e_1\frac{\partial f}{\partial x_1} + e_2\frac{\partial f}{\partial x_2} + e_3\frac{\partial f}{\partial x_3},$$

$$f\overline{\partial} = \frac{\partial f}{\partial x_0} + \frac{\partial f}{\partial x_1}e_1 + \frac{\partial f}{\partial x_2}e_2 + \frac{\partial f}{\partial x_3}e_3,$$

and the nth power can be written as follows:

$$\overline{\partial}_x^n = \sum_{k=0}^{n}\binom{n}{k}\left(\frac{\partial}{\partial x_0}\right)^{n-k}\left(\frac{\partial}{\partial x_1}e_1 + \frac{\partial}{\partial x_2}e_2 + \frac{\partial}{\partial x_3}e_3\right)^k,$$

where we note that the Moisil–Teodorescu operator

$$\frac{\partial}{\partial x_1}e_1 + \frac{\partial}{\partial x_2}e_2 + \frac{\partial}{\partial x_3}e_3$$

has a scalar square,

$$\left(\frac{\partial}{\partial x_1}e_1 + \frac{\partial}{\partial x_2}e_2 + \frac{\partial}{\partial x_3}e_3\right)^2 = -\left(\frac{\partial^2}{\partial x_1^2} + \frac{\partial^2}{\partial x_2^2} + \frac{\partial^2}{\partial x_3^2}\right),$$

which will make the components of its powers in each dimension easy to calculate.

How can we define fractional powers $\overline{\partial}_x^n$ of the quaternionic d-bar derivative? An answer is given by the following definition.

Definition 2.1 The fractional quaternionic d-bar derivative of a suitable quaternion-valued function $f(x) = f(x_0 + e_1x_1 + e_2x_2 + e_3x_3)$ with respect to x is defined in the following way:

$$\begin{aligned}
{}^{RL}_{c}\overline{\partial}_x^{\alpha} f &= \sum_{k=0}^{\infty} \binom{\alpha}{k} {}^{RL}_{c}D_{x_0}^{\alpha-k} \left(\frac{\partial}{\partial x_1}e_1 + \frac{\partial}{\partial x_2}e_2 + \frac{\partial}{\partial x_3}e_3\right)^k f \\
&= \sum_{\ell=0}^{\infty} (-1)^{\ell} \binom{\alpha}{2\ell} {}^{RL}_{c}D_{x_0}^{\alpha-2\ell} \left(\frac{\partial^2}{\partial x_1^2} + \frac{\partial^2}{\partial x_2^2} + \frac{\partial^2}{\partial x_3^2}\right)^{\ell} f \\
&\quad + \sum_{\ell=0}^{\infty} (-1)^{\ell} \binom{\alpha}{2\ell+1} {}^{RL}_{c}D_{x_0}^{\alpha-2\ell-1} \\
&\qquad \left(\frac{\partial^2}{\partial x_1^2} + \frac{\partial^2}{\partial x_2^2} + \frac{\partial^2}{\partial x_3^2}\right)^{\ell} \left(e_1\frac{\partial f}{\partial x_1} + e_2\frac{\partial f}{\partial x_2} + e_3\frac{\partial f}{\partial x_3}\right),
\end{aligned}$$

and likewise

$$\begin{aligned}
f\, {}^{RL}_{c}\overline{\partial}_x^{\alpha} &= \sum_{\ell=0}^{\infty} (-1)^{\ell} \binom{\alpha}{2\ell} {}^{RL}_{c}D_{x_0}^{\alpha-2\ell} \left(\frac{\partial^2}{\partial x_1^2} + \frac{\partial^2}{\partial x_2^2} + \frac{\partial^2}{\partial x_3^2}\right)^{\ell} f \\
&\quad + \sum_{\ell=0}^{\infty} (-1)^{\ell} \binom{\alpha}{2\ell+1} {}^{RL}_{c}D_{x_0}^{\alpha-2\ell-1} \left(\frac{\partial^2}{\partial x_1^2} + \frac{\partial^2}{\partial x_2^2} + \frac{\partial^2}{\partial x_3^2}\right)^{\ell} \\
&\qquad \left(\frac{\partial f}{\partial x_1}e_1 + \frac{\partial f}{\partial x_2}e_2 + \frac{\partial f}{\partial x_3}e_3\right).
\end{aligned}$$

Here, we use a constant of differintegration $c \in \mathbb{R}$ and a fractional order $\alpha \in \mathbb{R}$, and we make use of the Riemann–Liouville fractional differintegral ${}^{RL}_{c}D_{x_0}^{\beta}$ with respect to x_0 (per standard conventions [9], this is interpreted as a fractional derivative if $\operatorname{Re}\beta \geq 0$ and a fractional integral if $\operatorname{Re}\beta < 0$).

Our definition makes sense from the viewpoint of the binomial theorem, interpreting a fractional d-bar derivative formally as a fractional power of a sum of derivative operators. But, does it make sense as a fractional quaternionic derivative? We must find out whether standard properties of fractional derivatives are preserved as expected, and also whether standard properties of quaternionic derivatives are preserved as expected, in this new setting that combines two different mathematical fields.

3 Composition Results

Among the most important types of results in fractional calculus are composition results: what happens when we combine two fractional integral or derivative operators together? In the standard Riemann–Liouville fractional calculus [9], the answer is that a semigroup property is observed if we take an ordinary (non-fractional) derivative of any fractional differintegral or if we take any fractional differintegral of a fractional integral (derivative with negative order), but not in general. Here, we verify that the same holds true for our new operator.

Theorem 3.1 *For $\alpha \in \mathbb{R}$ and any suitable quaternionic function f, the following hold:*

$$\overline{\partial}\left({}^{RL}_{\ c}\overline{\partial}^{\alpha}_{x} f\right) = {}^{RL}_{\ c}\overline{\partial}^{\alpha+1}_{x} f, \qquad f\,{}^{RL}_{\ c}\overline{\partial}^{\alpha+1}_{x} = \left(f\,{}^{RL}_{\ c}\overline{\partial}^{\alpha}_{x}\right)\overline{\partial}.$$

Corollary 3.2 *For $\alpha \in \mathbb{R}$ and $n \in \mathbb{N}$ and any suitable quaternionic function f, the following hold:*

$$\overline{\partial}^{n}\left({}^{RL}_{\ c}\overline{\partial}^{\alpha}_{x} f\right) = {}^{RL}_{\ c}\overline{\partial}^{\alpha+n}_{x} f, \qquad f\,{}^{RL}_{\ c}\overline{\partial}^{\alpha+n}_{x} = \left(f\,{}^{RL}_{\ c}\overline{\partial}^{\alpha}_{x}\right)\overline{\partial}^{n}.$$

Theorem 3.3 *For $\alpha, \beta \in \mathbb{R}$ with $\beta < 0$, and for any suitable quaternionic function f, the following hold:*

$${}^{RL}_{\ c}\overline{\partial}^{\alpha}_{x}\left({}^{RL}_{\ c}\overline{\partial}^{\beta}_{x} f\right) = {}^{RL}_{\ c}\overline{\partial}^{\alpha+\beta}_{x} f, \qquad f\,{}^{RL}_{\ c}\overline{\partial}^{\alpha+\beta}_{x} = \left(f\,{}^{RL}_{\ c}\overline{\partial}^{\beta}_{x}\right){}^{RL}_{\ c}\overline{\partial}^{\alpha}_{x}.$$

The above results all show cases where a semigroup property is valid for the fractional quaternionic d-bar operator. To verify that it is not universally valid when $\beta > 0$, we also include the following result which is for $\beta = 1$.

Theorem 3.4 *For $\alpha \in \mathbb{R}$ and any suitable quaternionic function f, the following hold:*

$$
{}^{RL}_{c}\overline{\partial}^{\alpha}_{x}(\overline{\partial} f)(x) = {}^{RL}_{c}\overline{\partial}^{\alpha+1}_{x} f(x)
$$

$$
-\sum_{k=0}^{\infty}\binom{\alpha}{k}\frac{(x_0-c)^{k-\alpha-1}}{\Gamma(k-\alpha)}\left(\frac{\partial}{\partial x_1}e_1+\frac{\partial}{\partial x_2}e_2+\frac{\partial}{\partial x_3}e_3\right)^k f(c+e_1x_1+e_2x_2+e_3x_3),
$$

$$
\left((f\overline{\partial})\,{}^{RL}_{c}\overline{\partial}^{\alpha}_{x}\right)(x) = \left(f\,{}^{RL}_{c}\overline{\partial}^{\alpha+1}_{x}\right)(x)
$$

$$
-\sum_{k=0}^{\infty}\binom{\alpha}{k}\frac{(x_0-c)^{k-\alpha-1}}{\Gamma(k-\alpha)}f(c+e_1x_1+e_2x_2+e_3x_3)\left(\frac{\partial}{\partial x_1}e_1+\frac{\partial}{\partial x_2}e_2+\frac{\partial}{\partial x_3}e_3\right)^k.
$$

The proofs of all these results will be presented, along with suitable function spaces for f, in a forthcoming full paper.

4 Fueter Variables and Polynomials

A quaternionic function is said to be left-holomorphic if $\overline{\partial} f = 0$ or right-holomorphic if $f\overline{\partial} = 0$. (Here we use the "left" and "right" convention according to [5], although some other works [11] use the opposite convention.) These are suitable definitions for holomorphicity of quaternionic functions, but we must define variables and polynomials in a non-obvious way in order to guarantee their holomorphicity. The simple function $f(x) = x$ is neither left-holomorphic nor right-holomorphic, which led to the introduction of the Fueter variables

$$
z_k(x) = x_k - x_0 e_k
$$

for $k = 1, 2, 3$. These functions of x are both left-holomorphic and right-holomorphic [5], and they can be used to construct Fueter polynomials of any order as follows.

Definition 4.1 (Fueter polynomials) Let $\mathbf{k} = (k_1, \ldots, k_n)$ be a multi-index where each k_i is a non-negative integer. The degree of this multi-index and the corresponding Fueter term are defined as

$$
k = |\mathbf{k}| = \sum_{i=1}^{n} k_i \quad \text{and} \quad z_{\mathbf{k}} = z_1^{\mathbf{k_1}} \ldots z_n^{\mathbf{k_n}}.
$$

The Fueter polynomial $P_{\mathbf{k}}(x)$, for any $x \in \mathbb{H}$, is defined as follows.

- If there exist any negative k_i, then we define $P_{\mathbf{k}}(x) := 0$.
- If $\mathbf{k} = (0, \ldots, 0)$, shortly denoted as $\mathbf{k} = \mathbf{0}$, then we define $P_{\mathbf{0}}(x) := 1$.

- For $\mathbf{k}$ with $|\mathbf{k}| > 0$, the Fueter polynomial can be defined in the following way:

$$P_{\mathbf{k}}(x) := \frac{1}{k!} \sum_{\sigma \in perm(k)} \sigma\left(z_{\mathbf{k}}\right),$$

where $perm(k)$ is the permutation group on k elements.

The Fueter polynomials are necessary to define in order to have a quaternionic version of polynomials which are holomorphic. One of the important results concerning them [5, Theorem 6.2] is that every Fueter polynomial is both left-holomorphic and right-holomorphic: we have

$$\overline{\partial} P_{\mathbf{k}}(x) = P_{\mathbf{k}}(x)\overline{\partial} = 0$$

for any multi-index $\mathbf{k}$.

A natural question about the new fractional quaternionic d-bar derivative, then, is what happens when it is applied to Fueter polynomials. Note that we do not necessarily expect the answer to be zero: functions whose nth-order derivatives are always zero often have non-zero fractional derivatives, such as constant functions under the classical derivative and complex analytic functions under the d-bar derivative [1]. Here, some elementary calculation yields the following results:

$$\begin{aligned}
{}^{RL}_{c}\overline{\partial}^{\alpha}_{x}(x_1 - x_0 e_1) = (x_1 - x_0 e_1)\,{}^{RL}_{c}\overline{\partial}^{\alpha}_{x} &= \frac{(x_0 - c)^{-\alpha}}{\Gamma(1-\alpha)}(x_1 - x_0 e_1), \\
{}^{RL}_{c}\overline{\partial}^{\alpha}_{x}(x_2 - x_0 e_2) = (x_2 - x_0 e_2)\,{}^{RL}_{c}\overline{\partial}^{\alpha}_{x} &= \frac{(x_0 - c)^{-\alpha}}{\Gamma(1-\alpha)}(x_2 - x_0 e_2), \\
{}^{RL}_{c}\overline{\partial}^{\alpha}_{x}(x_3 - x_0 e_3) = (x_3 - x_0 e_3)\,{}^{RL}_{c}\overline{\partial}^{\alpha}_{x} &= \frac{(x_0 - c)^{-\alpha}}{\Gamma(1-\alpha)}(x_3 - x_0 e_3),
\end{aligned}$$

which we summarise in a theorem as follows.

Theorem 4.2 *The Fueter variables* $z_k = x_k - x_0 e_k$ *in* $\mathbb{H}$ *have fractional d-bar derivatives given by*

$$ {}^{RL}_{c}\overline{\partial}^{\alpha}_{x}(z_k) = (z_k)\,{}^{RL}_{c}\overline{\partial}^{\alpha}_{x} = \frac{(x_0 - c)^{-\alpha}}{\Gamma(1-\alpha)} z_k,$$

and therefore, as a particular case,

$$ {}^{RL}_{0}\overline{\partial}^{\alpha}_{x}(z_k) = (z_k)\,{}^{RL}_{0}\overline{\partial}^{\alpha}_{x} = \frac{x_0^{-\alpha}}{\Gamma(1-\alpha)} z_k,$$

valid for $k = 1, 2, 3$ *and any* $\alpha \in \mathbb{R}$.

We remark that the above result is natural, given that the Riemann–Liouville derivative of a constant function in classical fractional calculus is

$$ {}^{RL}_{c}D^{\alpha}_{x}(k) = \frac{(x-c)^{-\alpha}}{\Gamma(1-\alpha)}k $$

and we have a similar result [1, Example 3.8] for the fractional d-bar derivative of a complex analytic function.

More generally, what about the Fueter polynomials? Using the notation $\boldsymbol{\varepsilon}_1 = (1, 0, 0)$ and $\boldsymbol{\varepsilon}_2 = (0, 1, 0)$ and $\boldsymbol{\varepsilon}_3 = (0, 0, 1)$, we have some examples as follows.

Example 4.3 The Fueter polynomial $P_{\mathbf{k}}$ with $|\mathbf{k}| = 0$ is just the constant function 1, and its fractional d-bar derivatives are given as follows for any $\alpha \in \mathbb{R}$:

$$ {}^{RL}_{c}\overline{\partial}^{\alpha}_{x}(1) = (1)\,{}^{RL}_{c}\overline{\partial}^{\alpha}_{x} = \frac{(x_0-c)^{-\alpha}}{\Gamma(1-\alpha)}. $$

Example 4.4 The Fueter polynomials $P_{\mathbf{k}}$ with $|\mathbf{k}| = 1$ are just the Fueter variables: $P_{\boldsymbol{\varepsilon}_i} = z_i$ for $i = 1, 2, 3$. Their fractional d-bar derivatives are already given in Theorem 4.2 above.

Example 4.5 The Fueter polynomials $P_{\mathbf{k}}$ with $|\mathbf{k}| = 2$ come in two types: firstly,

$$ P_{2\boldsymbol{\varepsilon}_i} = z_i^2 $$

for $i = 1, 2, 3$, with for any $\alpha \in \mathbb{R}$

$$ {}^{RL}_{c}\overline{\partial}^{\alpha}_{x} P_{2\boldsymbol{\varepsilon}_i}(x) = \left(P_{2\boldsymbol{\varepsilon}_i}\,{}^{RL}_{c}\overline{\partial}^{\alpha}_{x}\right)(x) = \frac{(x_0-c)^{-\alpha}}{\Gamma(1-\alpha)}P_{2\boldsymbol{\varepsilon}_i}(x), $$

and secondly,

$$ P_{\boldsymbol{\varepsilon}_i+\boldsymbol{\varepsilon}_j} = \frac{1}{2}(z_i z_j + z_j z_i) $$

for $i < j$, with for any $\alpha \in \mathbb{R}$

$$ {}^{RL}_{c}\overline{\partial}^{\alpha}_{x} P_{\boldsymbol{\varepsilon}_i+\boldsymbol{\varepsilon}_j}(x) = \left(P_{\boldsymbol{\varepsilon}_i+\boldsymbol{\varepsilon}_j}\,{}^{RL}_{c}\overline{\partial}^{\alpha}_{x}\right)(x) = \frac{(x_0-c)^{-\alpha}}{\Gamma(1-\alpha)}P_{\boldsymbol{\varepsilon}_i+\boldsymbol{\varepsilon}_j}(x). $$

Example 4.6 If $\mathbf{k} = (1, 2, 0)$, then

$$ P_{\mathbf{k}}(x) = \frac{1}{3}\left(z_1 z_2{}^2 + z_2 z_1 z_2 + z_2{}^2 z_1\right), $$

and we have

$$ {}^{RL}_{\ c}\overline{\partial}_x^{-\alpha} P_{\mathbf{k}}(x) \neq \frac{(x_0-c)^{-\alpha}}{\Gamma(1-\alpha)} P_{\mathbf{k}}(x) \neq \left(P_{\mathbf{k}}\ {}^{RL}_{\ c}\overline{\partial}_x^{-\alpha}\right)(x). $$

So the pattern observed for low values of $|\mathbf{k}|$ does not hold true in general.

By using Theorem 3.4 and classical results on Fueter polynomials [5, Theorem 6.2], we were able to prove the following result.

Theorem 4.7 *For any $i = 1, 2, 3$ and $n \in \mathbb{N}$, we have*

$$ P_{n\boldsymbol{\varepsilon}_i}(x) = (x_i - x_0 e_i)^n , $$

and

$$ {}^{RL}_{\ c}\overline{\partial}_x^{-\alpha} P_{n\boldsymbol{\varepsilon}_i}(x) = \left(P_{n\boldsymbol{\varepsilon}_i}\ {}^{RL}_{\ c}\overline{\partial}_x^{-\alpha}\right)(x) = \frac{(x_0-c)^{-\alpha}}{\Gamma(1-\alpha)} P_{n\boldsymbol{\varepsilon}_i}(x). $$

This provides at least one infinite family of Fueter polynomials with very neat and natural formulae for their fractional d-bar derivatives.

The proofs of all the above results will be presented in a forthcoming full paper.

Acknowledgments This research is financially supported by Eastern Mediterranean University via a BAP-C grant with project number BAPC-04-22-03. The first two authors are also grateful to the Ghent Analysis & PDE group for hosting them during the period this research began.

References

1. A. Fernandez, C. Bouzouina, Fractionalisation of complex d-bar derivatives. Complex Variables Elliptic Equ. **66**(3), 437–475 (2021)
2. A.S. Fokas, D.A. Pinotsis, Quaternions, evaluation of integrals and boundary value problems. Comput. Methods Funct. Theor. **7**, 443–476 (2007)
3. G.R. Franssens, Introducing Clifford analysis as the natural tool for electromagnetic research, in *Proceedings of the Progress in Electromagnetics Research Symposium*, Moscow (2012)
4. G. Frobenius, Über lineare Substitutionen und bilineare Formen, J. für die Reine und Angew. Math. **84**, 1–63 (1878)
5. K. Gürlebeck, K. Habetha, W. Sprößig, *Holomorphic Functions in the Plane and n-Dimensional Space* (Birkhäuser, Berlin, 2008)
6. U. Kähler, N. Vieira, Fractional Clifford analysis, in *Hypercomplex Analysis: New Perspectives and Applications*, ed. by S. Bernstein, U. Kähler, I. Sabadini, F. Sommen (Springer International Publishing, Cham, 2014)
7. P. Lounesto, *Clifford Algebras and Spinors*, 2nd edn. (Cambridge University Press, Cambridge, 2001)
8. P.A. Nekrassov, On general differentiation. Mat. Sb. **14**(1), 45–168 (1888)
9. S.G. Samko, A.A. Kilbas, O.I. Marichev, *Fractional Integrals and Derivatives: Theory and Applications* (Gordon and Breach, Yverdon, 1993)

10. H.G. Sun, Y. Zhang, D. Baleanu, W. Chen, Y.Q. Chen, A new collection of real world applications of fractional calculus in science and engineering. Commun. Nonlinear Sci. Numer. Simul. **64**, 213–231 (2018)
11. D. Xu, C. Jahanchahi, C.C. Took, D.P. Mandic, Enabling quaternion derivatives: the generalized HR calculus. R. Soc. Open Sci. **2**(8), 150255 (2015)

A Note on Homogeneous Geometry of G_2-Type Real Flag Manifolds

Brian Grajales and Julieth Saavedra

Abstract In this note, we describe the invariant metrics on real flag manifolds of the split real form of the exceptional Lie group G_2, starting from the decomposition of the isotropy representation into irreducible invariant subspaces. This description establishes the basic pillars for the study of Riemannian geometry and homogeneous geometric flows on these manifolds.

2020 Mathematics Subject Classification 22F30, 53C30

1 Introduction

The exceptional Lie group G_2 is a mathematical object that has been the subject of intense study in the field of mathematics and physics. The study of its properties has important applications in several areas of science, including differential geometry, particle physics, and string theory. In particular, G_2 plays a central role in the theory of compact, seven-dimensional manifolds with holonomy G_2. These manifolds are of great interest to mathematicians and physicists, as they arise naturally in string theory and $M-$theory, which are two of the most important frameworks for studying the fundamental laws of physics. This note is dedicated to present some geometric issues of real flag manifolds of type G_2, which are homogeneous spaces of the split real form of G_2.

The study of the geometry and topology of homogeneous spaces is justified by its implementation in different branches of mathematics and physics. For real flag manifolds, which are quotients of non-compact connected Lie groups whose Lie algebra is a split real form of a complex Lie group by a parabolic subgroup,

B. Grajales (✉)
Departamento de Matemática, IMECC-Unicamp, Campinas, Brazil

J. Saavedra
Departamento de Matemática, Centro de Ciências, Fortaleza, Brazil

M. Ruzhansky, B. Torebek (eds.), *Extended Abstracts MWCAPDE 2023*, Research Perspectives Ghent Analysis and PDE Center 1,
https://doi.org/10.1007/978-3-031-41665-1_4

many topological aspects have been explored by several authors with different tools. For instance, a description of its mod 2 and integral homologies was obtained in [2] and [5]; its fundamental groups were described by Wiggerman in [11], orientability in [9], and different types of cohomologies in [1, 6, 7]. On geometric issues, the literature is relatively shorter. In fact, we had to wait until 2015 for a description of the isotropy representation of a real flag manifold [8]. This description is fundamental in order to understand and develop invariant Riemannian geometry of real flag manifolds. Among contributions to a better understanding of real flag manifolds, we point out a complete parametrization of Riemannian homogeneous metrics and a classification of geodesic orbit metrics for real flag manifolds of classical Lie algebras in [4] and the existence and description of Einstein invariant metrics when the isotropy representation decomposes into two or three irreducible subrepresentations [3], also for the classical case. In this note we will present the homogeneous space structure of G_2−type real flag manifolds and a description of the invariant metrics on these spaces.

2 Preliminaries

2.1 *A Model for* $\mathfrak{g}_2$

Let $\mathfrak{g}_2$ be the split real form of the smallest exceptional simple Lie algebra. We shall show the construction of a model of $\mathfrak{g}_2$ giving in [10, Section 8.4.1.].

Consider the Lie algebra $\mathfrak{sl}(3)$ of real traceless 3×3 matrices and the Euclidean three-dimensional space $\mathbb{R}^3$. Let $\{e_1, e_2, e_3\}$ be the canonical basis of $\mathbb{R}^3$ and $\{\epsilon_1, \epsilon_2, \epsilon_3\}$ be its dual basis. The 3rd exterior power $\bigwedge^3(\mathbb{R}^3)$ of $\mathbb{R}^3$ is isomorphic to $\mathbb{R}$ via

$$\Psi : c \in \mathbb{R} \mapsto c\nu \in \bigwedge^3(\mathbb{R}^3), \text{ where } \nu := e_1 \wedge e_2 \wedge e_3.$$

The map

$$T : \bigwedge^2(\mathbb{R}^3) \longrightarrow (\mathbb{R}^3)^*, \ T(u \wedge v)(w) := \Psi^{-1}(u \wedge v \wedge w)$$

and the map

$$S : \bigwedge^2(\mathbb{R}^3)^* \longrightarrow \mathbb{R}^3,$$

defined implicitly by the formula

$$\alpha \wedge \beta \wedge \gamma = \gamma(S(\alpha \wedge \beta))\nu^*, \text{ where } \nu^* := \epsilon_1 \wedge \epsilon_2 \wedge \epsilon_3, \text{ and } \alpha, \beta, \gamma \in (\mathbb{R}^3)^*,$$

are linear isomorphism.

The simple Lie algebra $\mathfrak{g}_2$ can be constructed over the vector space

$$\mathfrak{g} := \mathfrak{sl}(3) \oplus \mathbb{R}^3 \oplus \left(\mathbb{R}^3\right)^*$$

by defining the Lie brackets as follows:

- The Lie bracket of two matrices $X, Y \in \mathfrak{sl}(3)$ is given by

$$[X, Y] := XY - YX \in \mathfrak{sl}(3).$$

- If $X \in \mathfrak{sl}(3)$ and $v \in \mathbb{R}^3$, then

$$[X, v] := Xv \in \mathbb{R}^3.$$

- If $X \in \mathfrak{sl}(3)$ and $\alpha \in \left(\mathbb{R}^3\right)^*$, then

$$[X, \alpha] := -\alpha \circ X \in \left(\mathbb{R}^3\right)^*.$$

- If $u, v \in \mathbb{R}^3$ then

$$[u, v] := -\frac{4}{3} T(u \wedge v) \in \left(\mathbb{R}^3\right)^*.$$

- If $\alpha, \beta \in \left(\mathbb{R}^3\right)^*$, then

$$[\alpha, \beta] := \frac{4}{3} S(\alpha \wedge \beta) \in \mathbb{R}^3.$$

- If $v = \sum_{i=1}^{3} v^i e_i \in \mathbb{R}^3$ and $\alpha = \sum_{j=1}^{3} \alpha^j \epsilon_j \in \left(\mathbb{R}^3\right)^*$, then

$$[v, \alpha] := (v^i \alpha^j)_{3\times 3} - \frac{1}{3}\alpha(v)\mathrm{I}_3 \in \mathfrak{sl}(3),$$

  where I_3 is the identity matrix of order 3.
- For generic elements $X = X_1 + u_1 + \alpha_1$, $Y = X_2 + u_2 + \alpha_2 \in \mathfrak{g}$, $X_j \in \mathfrak{sl}(3)$, $u_j \in \mathbb{R}^3$, $\alpha_j \in \left(\mathbb{R}^3\right)^*$, $j = 1, 2$; define the Lie bracket $[X, Y]$ extending the rules above by linearity and skew-symmetry.

Proposition 2.1 *The vector space* $\mathfrak{g} = \mathfrak{sl}(3) \oplus \mathbb{R}^3 \oplus \left(\mathbb{R}^3\right)^*$ *endowed with the Lie bracket* $[\cdot, \cdot]$ *defined above is a Lie algebra isomorphic to* $\mathfrak{g}_2$.

2.2 A Root System

The subalgebra

$$\mathfrak{h} = \{A \in \mathfrak{sl}(3) : A \text{ is diagonal}\}$$

is a Cartan subalgebra of $\mathfrak{g}_2$, and its corresponding root system is

$$\Pi = \left\{\lambda_i - \lambda_j : 1 \leq i \neq j \leq 3\right\} \cup \{\pm\lambda_i : 1 \leq i \leq 3\},$$

where $\lambda_i(\operatorname{diag}(a_1, a_2, a_3)) := a_i$, $i = 1, 2, 3$. The root spaces are

$$(\mathfrak{g}_2)_{\lambda_i - \lambda_j} = \operatorname{span}\left\{E_{ij} := \left(\gamma_{kl}^{ij}\right)_{3\times 3}\right\}, \quad \text{where } \gamma_{kl}^{ij} := \delta_i^k \delta_l^j,$$

$$(\mathfrak{g}_2)_{\lambda_i} = \operatorname{span}\{e_i\}, \text{ and}$$

$$(\mathfrak{g}_2)_{-\lambda_i} = \operatorname{span}\{\epsilon_i\}.$$

The set of positive roots is $\Pi^+ = \left\{\lambda_i - \lambda_j : 1 \leq i < j \leq 3\right\} \cup \{\lambda_i : 1 \leq i \leq 3\}$, and a set of simple roots is given by $\Sigma = \{\alpha_1 := \lambda_1 - \lambda_2, \alpha_2 := \lambda_2\}$.

2.3 Real Flag Manifolds of $\mathfrak{g}_2$

Let G_2 be the connected simple Lie group whose Lie algebra is $\mathfrak{g}_2$. A real flag manifold of type $\mathfrak{g}_2$ is defined as the quotient space $\mathbb{F}_\Theta = G_2/P_\Theta$, where P_Θ is a parabolic subgroup of G_2. For any $\Theta \subseteq \Sigma$, we define the parabolic subalgebra $\mathfrak{p}_\Theta$ of $\mathfrak{g}_2$ and the corresponding parabolic subgroup P_Θ of G_2 as follows:

$$\mathfrak{p}_\Theta = \mathfrak{h} \oplus \bigoplus_{\alpha \in \Pi} (\mathfrak{g}_2)_\alpha \oplus \bigoplus_{\alpha \in \langle\Theta\rangle^-} (\mathfrak{g}_2)_\alpha$$

and

$$P_\Theta = \{a \in G_2 : \operatorname{Ad}(a)\mathfrak{p}_\Theta = \mathfrak{p}_\Theta\}.$$

Every parabolic subgroup P of G_2 can be expressed in the form $P = P_\Theta$, for some $\Theta \subseteq \Sigma$. Consequently, there exist only three real flag manifolds G_2/P_Θ corresponding to the subsets $\emptyset$, $\{\alpha_1\}$, and $\{\alpha_2\}$ of Σ.

The maximal compact subgroup K of G_2 acts transitively on $\mathbb{F}_\Theta$ with isotropy subgroup $K_\Theta = K \cap P_\Theta$; therefore a flag manifold $\mathbb{F}_\Theta = G_2/P_\Theta$ can also be seen as the homogeneous space K/K_Θ. If $\mathfrak{k}$ is the Lie algebra of K, then $\mathfrak{k} = \operatorname{span}\{X_1, X_2, X_3, Y_1, Y_2, Y_3\}$ where

$$X_1 := E_{21} - E_{12}, \; X_2 := E_{31} - E_{13}, \; X_3 := E_{32} - E_{23}, \; Y_i := e_i - \epsilon_i, \; i = 1, 2, 3$$

and these vectors satisfy the following bracket relations:

$$[X_1, X_2] = X_3, \quad [X_2, X_3] = X_1, \quad [X_3, Y_3] = -Y_2,$$

$$[X_1, X_3] = -X_2, [X_2, Y_1] = Y_3, \quad [Y_1, Y_2] = X_1 + \tfrac{4}{3}Y_3,$$

$$[X_1, Y_1] = Y_2, \quad [X_2, Y_3] = -Y_1, [Y_1, Y_3] = X_2 - \tfrac{4}{3}Y_2,$$

$$[X_1, Y_2] = -Y_1, \quad [X_3, Y_2] = Y_3, \quad [Y_2, Y_3] = X_3 + \tfrac{4}{3}Y_1.$$

3 Homogeneous Metrics

3.1 The Isotropy Representation

By compactness of K, the homogeneous space $\mathbb{F}_\Theta = K/K_\Theta$ is reductive; this means that there exists a decomposition $\mathfrak{k} = \mathfrak{k}_\Theta \oplus \mathfrak{m}_\Theta$ such that $\mathrm{Ad}(k)\mathfrak{m}_\Theta = \mathfrak{m}_\Theta$ for all $k \in K_\Theta$, so we can define the representation

$$\mathrm{Ad} : K_\Theta \longrightarrow \mathrm{GL}(\mathfrak{m}_\Theta). \tag{3.1}$$

This representation is equivalent to the isotropy representation of $\mathbb{F}_\Theta$, and it is completely reducible, that is, $\mathfrak{m}_\Theta$ can be decomposed as a direct sum of irreducible (not necessarily inequivalent) subrepresentations. The description of these subrepresentations and their equivalences is given in the following proposition proved by Patrão and San Martin in [8].

Proposition 3.1 *Let $\mathbb{F}_\Theta$ be a real flag manifold of $\mathfrak{g}_2$. Then the following statements hold:*

(*a*) *If $\Theta = \emptyset$, the representation* (3.1) *decomposes into the six one-dimensional linear spaces* $\mathrm{span}\{X_i\}, \mathrm{span}\{Y_i\}$, $i = 1, 2, 3$. *All these subspaces are obviously irreducible subrepresentations, and the equivalence classes among them are*

$$\{\mathrm{span}\{X_1\}, \mathrm{span}\{Y_3\}\},$$

$$\{\mathrm{span}\{X_2\}, \mathrm{span}\{Y_2\}\},$$

$$\{\mathrm{span}\{X_3\}, \mathrm{span}\{Y_1\}\}.$$

(*b*) *If* $\Theta = \{\alpha_1\}$, *the representation* (3.1) *decomposes into the irreducible subrepresentations*

$$\begin{aligned}&\text{span}\{Y_3\},\\&\text{span}\{X_2, X_3\},\\&\text{span}\{Y_2, Y_1\},\end{aligned}$$

where the two-dimensional subspaces are $K_\Theta-$*equivalent.*

(*c*) *If* $\Theta = \{\alpha_2\}$, *the representation* (3.1) *decomposes into the irreducible inequivalent subrepresentations*

$$\begin{aligned}&\text{span}\{X_2\},\\&\text{span}\left\{(\sqrt{13}-2)X_3 + 3Y_1, (\sqrt{13}-2)X_1 + 3Y_3\right\},\\&\text{span}\left\{(\sqrt{13}+2)X_3 - 3Y_1, (\sqrt{13}+2)X_1 - 3Y_3\right\}.\end{aligned}$$

3.2 $K-$Invariant Metrics

Fix the inner product $(\cdot,\cdot)$ on $\mathfrak{k}$ defined by

$$(X_i, X_j) = (Y_i, Y_j) = 2\delta_i^j \text{ and } (X_i, Y_j) = 0,\ i, j = 1, 2, 3.$$

Proposition 3.2 *The inner product* $(\cdot,\cdot)$ *above is* $\text{Ad}(K)-$*invariant and for every* $\Theta \subseteq \Sigma$, *the decomposition* $\mathfrak{k} = \mathfrak{k}_\Theta \oplus \mathfrak{m}_\Theta$ *is* $(\cdot,\cdot)-$*orthogonal.*

It is well-known that the set of $K-$invariant Riemannian metrics on $\mathbb{F}_\Theta$ is in one-to-one correspondence with the set of all $\text{Ad}(K_\Theta)-$invariant inner products on $\mathfrak{m}_\Theta$. On the other hand, if g is such an inner product, then there exists a $(\cdot,\cdot)-$self-adjoint positive linear operator $A_g : \mathfrak{m}_\Theta \to \mathfrak{m}_\Theta$ such that

$$A_g \circ \text{Ad}(k) = \text{Ad}(k) \circ A_g,\ \forall k \in K_\Theta \text{ and } g(X, Y) = (A_g X, Y), \forall X, Y \in \mathfrak{m}_\Theta.$$

We will say that A_g is the *metric operator associated to* g. Since the correspondence $g \mapsto A_g$ is bijective, we can understand $K-$invariant metrics on a flag $\mathbb{F}_\Theta$ as linear operators from $\mathfrak{m}_\Theta$ to $\mathfrak{m}_\Theta$ satisfying the conditions above. By fixing a basis of $\mathfrak{m}_\Theta$, these linear operators are nothing but matrices whose entries satisfy conditions derived from the properties of the operator.

Theorem 3.3 *Let* $\mathbb{F}_\Theta = K/K_\Theta$ *be a flag of* $\mathfrak{g}_2$ *and* g *be an* $\text{Ad}(K_\Theta)-$*invariant inner product on* $\mathfrak{m}_\Theta$.

(*a*) *If* $\Theta = \emptyset$, *then there exist positive real numbers* λ_i, $i = 1, 2, 3$ *and numbers* $b_i \in (-\lambda_i, \lambda_i), i = 1, 2, 3$ *such that the metric operator* A_g *associated with* g *can be written in the ordered basis* $\mathcal{B}_\emptyset = \{X_1, Y_3, X_2, Y_2, X_3, Y_1\}$ *as*

$$[A_g]_{\mathcal{B}_\emptyset} = \begin{pmatrix} \lambda_1 & b_1 & 0 & 0 & 0 & 0 \\ b_1 & \lambda_1 & 0 & 0 & 0 & 0 \\ 0 & 0 & \lambda_2 & b_2 & 0 & 0 \\ 0 & 0 & b_2 & \lambda_2 & 0 & 0 \\ 0 & 0 & 0 & 0 & \lambda_3 & b_3 \\ 0 & 0 & 0 & 0 & b_3 & \lambda_3 \end{pmatrix}.$$

(*b*) *If* $\Theta = \{\alpha_1\}$, *then there exist positive real numbers* λ_i, $i = 1, 2, 3$ *and numbers* $a, b \in (-\sqrt{\lambda_1\lambda_2}, \sqrt{\lambda_1\lambda_2}), i = 1, 2, 3$ *such that the metric operator* A_g *associated with* g *can be written in the ordered basis* $\mathcal{B}_{\{\alpha_1\}} = \{Y_3, X_2, X_3, Y_2, Y_1\}$ *as*

$$[A_g]_{\mathcal{B}_{\{\alpha_1\}}} = \begin{pmatrix} \lambda_1 & 0 & 0 & 0 & 0 \\ 0 & \lambda_2 & 0 & a & b \\ 0 & 0 & \lambda_2 & b & -a \\ 0 & a & b & \lambda_3 & 0 \\ 0 & b & -a & 0 & \lambda_3 \end{pmatrix}.$$

(*c*) *If* $\Theta = \{\alpha_2\}$, *then there exist positive real numbers* λ_i, $i = 1, 2, 3$ *such that the metric operator* A_g *associated with* g *can be written in the ordered basis* $\mathcal{B}_{\{\alpha_2\}} = \{X_2, (\sqrt{13} - 2)X_3 + 3Y_1, (\sqrt{13} - 2)X_1 + 3Y_3, (\sqrt{13} + 2)X_3 - 3Y_1, (\sqrt{13} + 2)X_1 - 3Y_3\}$ *as*

$$[A_g]_{\mathcal{B}_{\{\alpha_1\}}} = \begin{pmatrix} \lambda_1 & 0 & 0 & 0 & 0 \\ 0 & \lambda_2 & 0 & 0 & 0 \\ 0 & 0 & \lambda_2 & 0 & 0 \\ 0 & 0 & 0 & \lambda_3 & 0 \\ 0 & 0 & 0 & 0 & \lambda_3 \end{pmatrix}.$$

4 The Homogeneous Ricci Flow for $\Theta = \{\alpha_1\}$

The parametrization of homogeneous metrics provided by Theorem 3.3 allows us to deal with invariant geometry of real flag manifolds of type $\mathfrak{g}_2$ in a more practical way. As an example of this, we will see how the homogeneous Ricci flow equation for the flag $\mathbb{F}_{\{\alpha_2\}}$ can be presented as an autonomous ODE in $(\mathbb{R}^+)^3$. More precisely, we have the following theorem.

Theorem 4.1 *The homogeneous Ricci flow equation*

$$\frac{dg}{dt} = -2\mathrm{Ric}(g)$$

in the flag manifold $\mathbb{F}_{\{\alpha_2\}}$ *of* $\mathfrak{g}_2$ *is equivalent to the autonomous ODE*

$$\begin{pmatrix} \lambda_1' \\ \lambda_2' \\ \lambda_3' \end{pmatrix} = \begin{pmatrix} -\frac{1}{4}\left(\left(\frac{\lambda_1}{\lambda_2}\right)^2 + \left(\frac{\lambda_1}{\lambda_3}\right)^2\right) \\ \frac{\lambda_1}{4\lambda_2} - \frac{26}{26-2\sqrt{13}} \\ \frac{\lambda_1}{4\lambda_3} - \frac{26}{26+2\sqrt{13}} \end{pmatrix}, \ \lambda_1, \lambda_2, \lambda_3 > 0.$$

Acknowledgments Brian Grajales is supported by grant 2023/04083-0 (São Paulo Research Foundation FAPESP). Julieth Saavedra is supported by Instituto Serrapilheira.

References

1. V. Del Barco, L.A.B. San Martin, De Rham 2-Cohomology of Real Flag Manifolds. Symmetry Integrability Geom.: Methods Appl. (SIGMA) **15**, 23 (2019)
2. J.J. Duistermat, J.A.C. Kolk, V.S. Varadarajan, Functions, flows and oscillatory integral on flag manifolds. Compositio Math. **49**, 309–398 (1983)
3. B. Grajales, L. Grama, Invariant Einstein metrics on real flag manifolds with two or three isotropy summands. J. Geom. Phys. **176**, 104494 (2022)
4. B. Grajales, L. Grama, C.J.C. Negreiros, Geodesic Orbit Spaces in real flag manifolds. Commun. Anal. Geom. **28**(8), 1933–2003 (2020)
5. R.R. Kocherlakotar, Integral homology of real flag manifolds and loop spaces of symmetric spaces. Adv. Math. **110**, 1–46 (1995)
6. A.L. Mare, Equivariant cohomology of real flag manifolds. Differ. Geom. Appl. **24**, 223–229 (2006)
7. Á.K. Matszangosz, On the cohomology rings of real flag manifolds: Schubert cycles. Math. Ann. **381**, 1537–1588 (2021)
8. M. Patrão, L.A.B. San Martin, The isotropy representation of a real flag manifold: split real forms. Indag. Math. (N.S.) **26**(3), 547–579 (2015)
9. M. Patrão, L.A.B. San Martin, L.J. dos Santos, L. Seco, Orientability of vector bundles over real flag manifolds. Topology Appl. **159**(10–11), 2774–2786 (2012)
10. L.A.B. San Martin, *Algebras de Lie* (Editora Unicamp, 2010)
11. M. Wiggerman, The fundamental group of a real flag manifold. Indag. Mathem. **9**(1), 141–153 (1998)

Titchmarsh Theorems for Fourier Transforms of Hölder-Lipschitz Functions on Fundamental Domains of Lattices in $\mathbb{R}^d$

Arne Hendrickx

2000 Mathematics Subject Classification 30C65, 28A78, 26B35

1 Introduction

In 1948 Titchmarsh proved two theorems about Hölder-Lipschitz spaces on $\mathbb{R}$ in [6], which now bear his name. The first theorem is an extension of the Hausdorff-Young inequality. This inequality implies that $\widehat{f} \in L^{p'}$ if $f \in L^p$ for $1 \leq p \leq 2$. Titchmarsh extended the range of exponents β for which $\widehat{f} \in L^\beta$ under some Lipschitz-type condition on $f \in L^p$.

Theorem 1 ([6, Theorem 84]) *Suppose that $f \in L^p(\mathbb{R})$, for some $1 < p \leq 2$, satisfies the Lipschitz-type condition*

$$\int_{-\infty}^{\infty} |f(x+h) - f(x-h)|^p \, \mathrm{d}x = O\left(h^{\alpha p}\right) \quad as \quad h \to 0.$$

Then its Fourier transform $\widehat{f}$ belongs to $L^\beta(\mathbb{R})$ for

$$\frac{p}{p + \alpha p - 1} < \beta \leq \frac{p}{p-1}.$$

This research has been partially supported by Fonds Wetenschappelijk Onderzoek – Vlaanderen (FWO) with the PhD Fellowship Fundamental Research grant 1187323N

A. Hendrickx (✉)
Department of Mathematics: Analysis, Logic and Discrete Mathematics, Ghent University, Ghent, Belgium
e-mail: arnhendr.Hendrickx@UGent.be

M. Ruzhansky, B. Torebek (eds.), *Extended Abstracts MWCAPDE 2023*, Research Perspectives Ghent Analysis and PDE Center 1,
https://doi.org/10.1007/978-3-031-41665-1_5

The second theorem of Titchmarsh is a characterisation of L^2-functions satisfying a certain Lipschitz-type condition in terms of an asymptotic estimate for an integral over its Fourier transform.

Theorem 2 ([6, Theorem 85]) *Suppose* $f \in L^2(\mathbb{R})$ *and* $0 < \alpha < 1$. *Then the condition*

$$\int_{-\infty}^{\infty} |f(x+h) - f(x-h)|^2 \, \mathrm{d}x = O\left(|h|^{2\alpha}\right) \quad \text{as} \quad h \to 0$$

is equivalent with

$$\left(\int_{-\infty}^{-X} + \int_{X}^{\infty}\right) |\widehat{f}(\xi)|^2 \, \mathrm{d}\xi = O\left(X^{-2\alpha}\right) \quad \text{as} \quad X \to \infty.$$

These theorems have attracted the attention of several mathematicians, and there are several generalisations of them. For example, Younis considered Titchmarsh theorems on the torus (one- and two-dimensional cases) and on compact zero-dimensional groups in [7]. He studied Dini-Lipschitz functions on $\mathbb{R}^n$ in [8] and provided some hints to get the obtained results on $\mathbb{T}^n$. Titchmarsh theorems for these function spaces on compact homogeneous manifolds are examined in [1]. A more general version of the second Titchmarsh theorem via moduli of continuity is considered in [2]. Several other investigations have been made in the papers in the reference list.

My research focused on generalising the Titchmarsh theorems to the setting of fundamental domains of lattices. A fundamental domain of a lattice $L = A\mathbb{Z}^d$, where the generator matrix A is an invertible matrix, is a measurable set $\Omega \subseteq \mathbb{R}^d$ such that $\Omega + L = \mathbb{R}^d$ as a direct sum. It follows easily that Ω is homeomorphic to $A\mathbb{T}^d$. Fuglede's theorem [4, Sect. 6] ensures that $\{e^{2\pi i \kappa \cdot x} \mid \kappa \in L^\perp\}$ is an orthonormal basis for $L^2(\Omega)$, which we endow with the normalised measure $\mathrm{d}x / |\Omega|$, where the dual lattice is defined as $L^\perp := \left(A^\mathrm{T}\right)^{-1} \mathbb{Z}^d$. In this way one can define the Fourier transform $\mathcal{F}_\Omega : L^2(\Omega) \to \ell^2(L^\perp)$ by

$$\widehat{f}(\kappa) := \mathcal{F}_\Omega f(\kappa) := \frac{1}{|\Omega|} \int_\Omega f(x) \, e^{-2\pi i \kappa \cdot x} \, \mathrm{d}x$$

with inverse transform

$$\check{f}(x) := \mathcal{F}_\Omega^{-1} f(x) := \sum_{\kappa \in L^\perp} f(\kappa) \, e^{2\pi i \kappa \cdot x}.$$

A more detailed summary of the necessary concepts can be found in [5].

2 Titchmarsh Theorems for Hölder-Lipschitz Spaces

Our study of Titchmarsh theorems starts by defining appropriate Hölder-Lipschitz spaces on fundamental domains of lattices. The fundamental domain Ω lives in a higher dimensional Euclidean space $\mathbb{R}^d$, and there are multiple ways for defining Hölder-Lipschitz spaces. Maybe the most natural one that first comes to mind is the *additive* Hölder-Lipschitz space that is defined via a global translation in the following way.

Definition 1 Let $\alpha \in \mathbb{R}$ with $0 < \alpha \leq 1$, and let $1 < p \leq \infty$. We define the *Hölder-Lipschitz space* $\mathrm{Lip}_\Omega(\alpha; p)$ by

$$\mathrm{Lip}_\Omega(\alpha; p) := \left\{ f \in L^p(\Omega) : \|f(\cdot + h) - f(\cdot)\|_{L^p(\Omega)} = O\left(|h|^\alpha\right) \text{ as } h \to 0 \right\}.$$

However, Younis studied the *multiplicative* Hölder-Lipschitz spaces on the one- and two-dimensional torus [7], which arises from taking the composition of all componentwise translations as the defining translation operator. We should be careful to define componentwise translations in the right way; namely we need to take the transformation of the generator matrix A into account to avoid problems later on. Note that this is the natural componentwise translation on $\Omega \cong A\mathbb{T}^d$.

Definition 2 We define for any $1 \leq j \leq d$ and $h_j \in \mathbb{R}$ the difference operator $\Delta_j^{h_j}$ acting on a function $f \in L^p(\Omega)$ for some $1 \leq p \leq \infty$ by

$$\Delta_j^{h_j} f(x) := f\left(x + A\left(h_j e_j\right)\right) - f(x),$$

where e_j is the j-th element of the standard basis of $\mathbb{R}^d$.

With this set of componentwise translation operators, we are ready to define the multiplicative Hölder-Lipschitz spaces.

Definition 3 Let $\alpha_1, \ldots, \alpha_d \in \mathbb{R}$ with $0 < \alpha_1, \ldots, \alpha_d \leq 1$, and let $1 < p \leq \infty$. We define the (multiplicative) *Hölder-Lipschitz space* $\mathrm{Lip}_\Omega^M(\alpha_1, \ldots, \alpha_d; p)$ by

$$\begin{aligned}\mathrm{Lip}_\Omega^M(\alpha_1, \ldots, \alpha_d; p) := \Bigg\{ f \in L^p(\Omega) : &\left\| \Delta_1^{h_1} \ldots \Delta_d^{h_d} f \right\|_{L^p(\Omega)} \\ &= O\left(h_1^{\alpha_1} \ldots h_d^{\alpha_d}\right) \text{ as } h \to 0 \Bigg\}.\end{aligned}$$

We abbreviate $\mathrm{Lip}_\Omega^M(\alpha, \ldots, \alpha; p)$ as $\mathrm{Lip}_\Omega^M(\alpha; p)$.

Let us note a last important concept of the dual lattice $L^\perp$. While $\kappa_j := \kappa \cdot e_j$ for $1 \leq j \leq d$ is a set of coordinates for $\kappa \in L^\perp$, it is not the most natural one due to the deformation caused by the matrix A. Indeed, a more natural set of coordinates is given by $\kappa_j^A := \kappa \cdot Ae_j$. If we write $\kappa = A^\perp m$ for some $m \in \mathbb{Z}^d$, we find that

$\kappa_j^A = m_j$, so that it essentially uses the standard lattice $\mathbb{Z}^d$ for parameterising the lattice $L^\perp$.

2.1 Additive Case

Let us first explore the Titchmarsh theorems for additive Hölder-Lipschitz spaces. The following result can be obtained by using a strategy similar to the one in [7, Theorem 2.15] for the case of $\mathbb{R}^d$.

Theorem 3 *If* $f \in \mathrm{Lip}_\Omega(\alpha; p)$ *for some* $0 < \alpha \leq 1$ *and* $1 < p \leq 2$, *then* $\widehat{f} \in \ell^\beta(L^\perp)$ *for*

$$\frac{p}{p + \frac{\alpha}{d}p - 1} < \beta \leq p' = \frac{p}{p-1}.$$

Remark that this result becomes weaker as the dimension increases because the range for β gets smaller. It is known that the lower bound for β is sharp as it is shown in [9, page 243] for $d = 1$ and $L = \mathbb{Z}$ that the function

$$f(x) = \sum_{n=1}^{\infty} \frac{e^{in \log n}}{n^{\frac{1}{2}+\alpha}}$$

for $0 < \alpha < 1$ is in $\mathrm{Lip}_\mathbb{T}(\alpha; 2)$ but $\widehat{f} \notin \ell^{\frac{2}{2\alpha+1}}(\mathbb{Z})$.

Next, we investigate the second Titchmarsh theorem for additive Hölder-Lipschitz spaces. Its proof is actually a special case of the one of Theorem 14.

Theorem 4 *Let* $0 < \alpha < 1$ *and* $f \in L^2(\Omega)$. *Then* $f \in \mathrm{Lip}_\Omega(\alpha; 2)$ *if and only if*

$$\sum_{1 \leq |\kappa \cdot h|} |\widehat{f}(\kappa)|^2 = O\left(|h|^{2\alpha}\right).$$

The reason for using the summation range $\{\kappa \in L^\perp : 1 \leq |\kappa \cdot h|\}$ is to be able to use Jordan's inequality $2\,|\kappa \cdot h| \leq |\sin(\pi\kappa \cdot h)|$ in the proof.

2.2 Multiplicative Case

We now move on to a study of multiplicative Hölder-Lipschitz spaces. Because of their definition by componentwise translations, they can be treated via methods from the one-dimensional case. Firstly, we present the first Titchmarsh theorem, the proof of which is inspired by arguments in [7, Theorem 2.13].

Theorem 5 *If* $f \in \mathrm{Lip}_{\Omega}^{M}(\alpha; p)$ *for some* $0 < \alpha \leq 1$ *and* $1 < p \leq 2$, *then* $\widehat{f} \in \ell^{\beta}(L^{\perp})$ *for*

$$\frac{p}{p+\alpha p-1} < \beta \leq p' = \frac{p}{p-1}.$$

Notice that, contrary to the first Titchmarsh theorem for additive Hölder-Lipschitz spaces, this result for multiplicative Hölder-Lipschitz spaces is equally strong in all dimensions. The proof of this theorem relies on a generalisation of Duren's lemma [3, page 101] to higher dimensions.

Lemma 6 (Duren) *Let* $d \in \mathbb{N}$. *Consider a non-negative sequence* $c_{n_1,\dots,n_d} \geq 0$, *and let* $0 < a_n < b_n$ *for every* $1 \leq n \leq d$. *Then all the estimates of the following form are equivalent:*

$$\sum\nolimits_1 \cdots \sum\nolimits_d n_1^{\varepsilon_1 b_1} \dots n_d^{\varepsilon_d b_d} c_{n_1,\dots,n_d} = O\left(N_1^{a_1-(1-\varepsilon_1)b_1} \dots N_d^{a_d-(1-\varepsilon_d)b_d}\right),$$

where every sum $\sum_j$ *is of the form*

$$\sum_{n_j=1}^{N_j}, \quad \text{in which case we set } \varepsilon_j = 1,$$

or

$$\sum_{n_j \geq N_j}, \quad \text{in which case we set } \varepsilon_j = 0.$$

In particular, we have

$$\sum_{n_1=1}^{N_1} \cdots \sum_{n_d=1}^{N_d} n_1^{b_1} \dots n_d^{b_d} c_{n_1,\dots,n_d} = O\left(N_1^{a_1} \dots N_d^{a_d}\right)$$

if and only if

$$\sum_{n_1 \geq N_1} \cdots \sum_{n_d \geq N_d} c_{n_1,\dots,n_d} = O\left(N_1^{a_1-b_1} \dots N_d^{a_d-b_d}\right).$$

Now, we proceed to the statement of the second Titchmarsh theorem for multiplicative Hölder-Lipschitz spaces. The proof follows the same reasoning as [7, Theorem 2.19], which again relies on Duren's lemma.

Theorem 7 *Let* $0 < \alpha_1, \ldots, \alpha_d \leq 1$ *and* $f \in L^2(\Omega)$. *Then* $f \in \mathrm{Lip}_\Omega^M(\alpha_1, \ldots, \alpha_d; 2)$ *if and only if any estimate of the following form holds:*

$$\sum_{\kappa_1^A \in I_{\varepsilon_1}(N_1)} \cdots \sum_{\kappa_d^A \in I_{\varepsilon_d}(N_d)} \prod_{j=1}^{d} |\kappa_j^A|^{2\varepsilon_j} |\widehat{f}(\kappa)|^2 = O\left(N_1^{-2\alpha_1+2\varepsilon_1} \ldots N_d^{-2\alpha_d+2\varepsilon_d}\right), \tag{2.1}$$

where $\varepsilon_1, \ldots, \varepsilon_d \in \{0, 1\}$, $I_1(N) := [-N, N]$ *and* $I_0(N) := \mathbb{R} \setminus [-N, N]$ *for* $N \in \mathbb{N}$. *In particular, we have that* $f \in \mathrm{Lip}_\Omega^M(\alpha; 2)$ *if and only if*

$$\sum_{|\kappa_1^A| > N} \cdots \sum_{|\kappa_d^A| > N} |\widehat{f}(\kappa)|^2 = O\left(N^{-2\alpha_1} \ldots N^{-2\alpha_d}\right).$$

Notice that Duren's lemma guarantees that if one of the estimates (2.1) holds, then they actually all hold.

3 Titchmarsh Theorems for Dini-Lipschitz Spaces

Now that we have studied Titchmarsh theorems for Hölder-Lipschitz spaces, we can start to wonder if we can prove similar results under a Lipschitz-type condition with a different error bound. Hence, we start examining Dini-Lipschitz spaces. Again, we can consider an additive and a multiplicative case.

3.1 Additive Case

We can define additive Dini-Lipschitz spaces in a similar way as Hölder-Lipschitz spaces but note that our error bound is not globally defined on the positive real line now.

Definition 4 Let $0 < \alpha \leq 1$, $\beta \in \mathbb{R}$ and $1 < p \leq 2$. We define the additive Dini-Lipschitz space $\mathrm{DinLip}_\Omega(\alpha; \beta; p)$ by

$$\mathrm{DinLip}_\Omega(\alpha; \beta; p) := \left\{ f \in L^p(\Omega) : \|f(\cdot + h) - f(\cdot)\|_{L^p(\Omega)} = O\left(\frac{|h|^\alpha}{\log^\beta \frac{1}{|h|}}\right) \text{ as } h \to 0 \right\}.$$

We obtain the first Titchmarsh theorem for additive Dini-Lipschitz spaces in a similar way as Theorem 3. The difference with the corresponding result for additive

Hölder-Lipschitz spaces is that now β can attain the lower bound under a suitable condition.

Theorem 8 *If $f \in \mathrm{DinLip}_\Omega(\alpha;\beta;p)$ for some $0<\alpha\leq 1$, $\beta\in\mathbb{R}$ and $1<p\leq 2$, then $\widehat{f}\in\ell^\gamma(L^\perp)$ for*

$$\frac{p}{p+\frac{\alpha}{d}p-1}<\gamma\leq p'=\frac{p}{p-1}.$$

Moreover, if $\beta>0$, then we also have that $\widehat{f}\in\ell^\gamma(L^\perp)$ for

$$\gamma=\frac{p}{p+\frac{\alpha}{d}p-1}.$$

The second Titchmarsh theorem can be proven with the same strategy as Theorem 4. The only notable difficulty is that the error bound for the Lipschitz-type condition is not globally defined, but we can resolve this problem with a convenient extension procedure.

Theorem 9 *Let $0<\alpha<1$ and $\beta\in\mathbb{R}$. Then $f\in\mathrm{DinLip}_\Omega(\alpha;\beta;2)$ if and only if*

$$\sum_{1\leq|\kappa\cdot h|}|\widehat{f}(\kappa)|^2=O\left(\frac{|h|^{2\alpha}}{\log^{2\beta}\frac{1}{|h|}}\right).$$

3.2 *Multiplicative Case*

In the same way as before, we can define multiplicative Dini-Lipschitz spaces.

Definition 5 Let $0<\alpha\leq 1$, $\beta\in\mathbb{R}$ and $1<p\leq 2$. We define the multiplicative Dini-Lipschitz space $\mathrm{DinLip}_\Omega^M(\alpha;\beta;p)$ by

$$\mathrm{DinLip}_\Omega^M(\alpha;\beta;p):=\left\{f\in L^p(\Omega):\left\|\Delta_1^{h_1}\ldots\Delta_d^{h_d}f\right\|_{L^p(\Omega)}\right.$$
$$\left.=O\left(\frac{h_1^\alpha}{\log^\beta\frac{1}{h_1}}\cdots\frac{h_d^\alpha}{\log^\beta\frac{1}{h_d}}\right)\right\}.$$

Also for these spaces we are able to prove the second Titchmarsh theorem.

Theorem 10 *If $f\in\mathrm{DinLip}_\Omega^M(\alpha;\beta;p)$ for some $0<\alpha\leq 1$, $\beta\in\mathbb{R}$ and $1<p\leq 2$, then $\widehat{f}\in\ell^\gamma(L^\perp)$ for*

$$\frac{p}{p+\alpha p-1}<\gamma\leq p'=\frac{p}{p-1}.$$

Moreover, if $\beta > \alpha + \frac{1}{p'}$*, then we also have that* $\widehat{f} \in \ell^{\gamma}(L^{\perp})$ *for* $\gamma = \frac{p}{p+\alpha p-1}$.

The proof of this theorem is similar to that of Theorem 5. Since that proof relies on Duren's lemma, we need a similar result for Dini-Lipschitz functions. This result and its proof are based on [1, Lemma 4.1].

Lemma 11 (Duren-type lemma for Dini-Lipschitz functions) *Let* $d \in \mathbb{N}$. *Consider a non-negative sequence* $c_{n_1,\ldots,n_d} \geq 0$, *and let* $\beta_1, \ldots, \beta_d \in \mathbb{R}$ *and* $0 < a_n < b_n$ *for every* $1 \leq n \leq d$. *Then all the estimates of the following form are equivalent:*

$$\sum\nolimits_1 \cdots \sum\nolimits_d n_1^{\varepsilon_1 b_1} \ldots n_d^{\varepsilon_d b_d} c_{n_1,\ldots,n_d} = O\left(\frac{N_1^{a_1-(1-\varepsilon_1)b_1}}{\log^{\beta_1} N_1} \cdots \frac{N_d^{a_d-(1-\varepsilon_d)b_d}}{\log^{\beta_d} N_d}\right),$$

where every sum $\sum_j$ *is of the form*

$$\sum_{n_j=1}^{N_j}, \quad \text{in which case we set } \varepsilon_j = 1,$$

or

$$\sum_{n_j \geq N_j}, \quad \text{in which case we set } \varepsilon_j = 0.$$

In particular, we have

$$\sum_{n_1=1}^{N_1} \cdots \sum_{n_d=1}^{N_d} n_1^{b_1} \ldots n_d^{b_d} c_{n_1,\ldots,n_d} = O\left(\frac{N_1^{a_1}}{\log^{\beta_1} N_1} \cdots \frac{N_d^{a_d}}{\log^{\beta_d} N_d}\right)$$

if and only if

$$\sum_{n_1 \geq N_1} \cdots \sum_{n_d \geq N_d} c_{n_1,\ldots,n_d} = O\left(\frac{N_1^{a_1-b_1}}{\log^{\beta_1} N_1} \cdots \frac{N_d^{a_d-b_d}}{\log^{\beta_d} N_d}\right).$$

Now, observe that in Theorem 10 the parameter β can attain its lower bound under a suitable condition, but note that this condition is more restrictive than in Theorem 8. This is because we used the following lemma.

Lemma 12 (Boundary case of Duren-type lemma for Dini-Lipschitz functions) *Let* $d \in \mathbb{N}$. *Consider a non-negative sequence* $c_{n_1,\ldots,n_d} \geq 0$, *and let* $\beta_1, \ldots, \beta_d > 1$ *and* $a_n > 0$ *for every* $1 \leq n \leq d$. *If*

$$\sum_{n_1=1}^{N_1} \cdots \sum_{n_d=1}^{N_d} n_1^{a_1} \ldots n_d^{a_d} c_{n_1,\ldots,n_d} = O\left(\frac{N_1^{a_1}}{\log^{\beta_1} N_1} \cdots \frac{N_d^{a_d}}{\log^{\beta_d} N_d}\right),$$

then

$$\sum_{n_1 \geq N_1} \cdots \sum_{n_d \geq N_d} c_{n_1,\ldots,n_d} = O\left(\frac{1}{\log^{\beta_1 - 1} N_1} \cdots \frac{1}{\log^{\beta_d - 1} N_d}\right).$$

Indeed, it is an interesting question if we can improve the former lemma in such a way that we have the error bound $O\left(\frac{1}{\log^{\beta_1} N_1} \cdots \frac{1}{\log^{\beta_d} N_d}\right)$ instead of $O\left(\frac{1}{\log^{\beta_1 - 1} N_1} \cdots \frac{1}{\log^{\beta_d - 1} N_d}\right)$ for the second sum. I leave this question open. This would directly improve Theorem 10. Since in the one-dimensional case additive and multiplicative Dini-Lipschitz spaces are the same and we already know that the weaker condition $\beta > 0$ is sufficient for additive Dini-Lipschitz spaces to ensure that the lower bound can be attained, this suggests that it might be possible to prove this for multiplicative Dini-Lipschitz spaces in higher dimensions.

We now turn to the second Titchmarsh theorem for multiplicative Dini-Lipschitz spaces, for which we define the following function space.

Definition 6 Let $0 < \alpha_1, \ldots, \alpha_d \leq 1$, $\beta_1, \ldots, \beta_d \in \mathbb{R}$ and $1 < p \leq 2$. We define the (multiplicative) *Dini-Lipschitz space* $\operatorname{DinLip}_\Omega^M(\alpha_1, \ldots, \alpha_d; \beta_1, \ldots, \beta_d; p)$ as the space of all functions $f \in L^p(\Omega)$ satisfying

$$\left\| \Delta_1^{h_1} \ldots \Delta_d^{h_d} f \right\|_{L^p(\Omega)} = O\left(\frac{h_1^{\alpha_1}}{\log^{\beta_1} \frac{1}{h_1}} \cdots \frac{h_d^{\alpha_d}}{\log^{\beta_d} \frac{1}{h_d}}\right) \quad \text{as } h = (h_1, \ldots, h_d) \to 0.$$

Again, the second Titchmarsh theorem can be obtained by using the methods from the multiplicative Hölder-Lipschitz spaces, namely Theorem 7, with the Duren-type Lemma 11 instead of Lemma 6.

Theorem 13 *Let* $0 < \alpha_1, \ldots, \alpha_d \leq 1$, $\beta_1, \ldots, \beta_d \in \mathbb{R}$, *and let* $f \in L^2(\Omega)$. *Then* $f \in \operatorname{DinLip}_\Omega^M(\alpha_1, \ldots, \alpha_d; \beta_1, \ldots, \beta_d; 2)$ *if and only if any estimate of the following form holds:*

$$\sum_{\kappa_1^A \in I_{\varepsilon_1}(N_1)} \cdots \sum_{\kappa_d^A \in I_{\varepsilon_d}(N_d)} \prod_{j=1}^{d} |\kappa_j^A|^{2\varepsilon_j} |\widehat{f}(\kappa)|^2 = O\left(\frac{N_1^{-2\alpha_1 + 2\varepsilon_1}}{\log^{2\beta_1} N_1} \cdots \frac{N_d^{-2\alpha_d + 2\varepsilon_d}}{\log^{2\beta_d} N_d}\right).$$

4 Moduli of Continuity

After treating Hölder-Lipschitz and Dini-Lipschitz spaces, the question arises if we can consider a general class of error bounds for the Lipschitz-type conditions for which we can show the Titchmarsh theorems. This question has been considered in [2], for example. We will show that we can obtain the second Titchmarsh theorems if we take moduli of continuity as error bounds.

Definition 7 A function $\omega : I \ni 0 \to [0, \infty)$ defined on an interval $I \subseteq [0, \infty)$ is called a *modulus of continuity* if there exists some $\delta_0 > 0$ such that

(MOC1) $\omega(0) = 0$ and ω is continuous on $[0, \delta_0]$.
(MOC2) There is some $C \geq 1$ such that $\omega(t) \leq C\omega(s)$ for all $t, s \in [0, \delta_0]$ with $t \leq s$.
(MOC3) There is some $C \geq 1$ such that $\frac{\omega(t)}{t} \leq C\frac{\omega(s)}{s}$ for all $t, s \in (0, \delta_0]$ with $t \geq s$.

Moreover, we assume that ω satisfies the Zygmund-type condition

$$\int_0^t \frac{\omega(x)}{x}\, dx \leq C\omega(t) \text{ for all } t \in [0, \delta_0].$$

For example, in the case of Hölder-Lipschitz spaces the modulus of continuity is $\omega(t) = t^{2\alpha}$ with $0 < \alpha < 1/2$. Note that we cannot include the cases $1/2 \leq \alpha < 1$ because of the definition of moduli of continuity.

The proof of the following theorem is based on [2, Theorem 3.1]. This result is the analogue for the second Titchmarsh theorem for additive Hölder-Lipschitz and Dini-Lipschitz spaces.

Theorem 14 *Let ω be a modulus of continuity, $0 < \alpha < 2$ and $f \in L^2(\Omega)$. Then we have that*

$$\|f(\cdot + h) - f(\cdot)\|^2_{L^2(\Omega)} = O\big(\omega^\alpha(|h|)\big) \quad \textit{as} \quad h \to 0$$

if and only if

$$\sum_{\substack{\kappa \in L^\perp \\ 1 \leq |\kappa \cdot h|}} |\widehat{f}(\kappa)|^2 = O\big(\omega^\alpha(|h|)\big) \quad \textit{as} \quad h \to 0.$$

Note that the power α allows us to exactly recover the classical theorems for Hölder-Lipschitz and Dini-Lipschitz functions.

Finally, we can also get a general version of the second Titchmarsh theorem for multiplicative Lipschitz-type spaces by using methods from Theorem 7 and the general ideas for moduli of continuity from [2]. Note that we can now allow any positive power of the modulus of continuity in the error term instead of restricting the range to $(0, 2)$.

Theorem 15 *Let $\omega_1, \ldots, \omega_d$ be moduli of continuity, $\alpha_1, \ldots, \alpha_d > 0$ and let $f \in L^2(\Omega)$. Then we have*

$$\left\|\Delta_1^{h_1} \ldots \Delta_d^{h_d} f\right\|_{L^2(\Omega)} = O\big(\omega_1^{\alpha_1}(|h_1|) \ldots \omega_d^{\alpha_d}(|h_d|)\big) \quad \textit{as} \quad h = (h_1, \ldots, h_d) \to 0$$

if and only if every estimate of the following form holds:

$$\sum_{\kappa\in A^{\perp}R_{\varepsilon}\left(\frac{1}{h}\right)}\prod_{j=1}^{d}|\kappa_j^{A}|^{2\varepsilon_j}|\widehat{f}(\kappa)|^2 = O\left(\frac{\omega_1^{\alpha_1}(|h_1|)}{|h_1|^{2\varepsilon_1}}\cdots\frac{\omega_d^{\alpha_d}(|h_d|)}{|h_d|^{2\varepsilon_d}}\right) \quad as \quad h\to 0, \tag{4.1}$$

where $R_{\varepsilon}\left(\frac{1}{h}\right) := \prod_{j=1}^{d} I_{\varepsilon_j}(\frac{1}{h_j})$ *and we use the same notation as in Theorem* 7.

References

1. R. Daher, J. Delgado, M. Ruzhansky, Titchmarsh theorems for Fourier transforms of Hölder-Lipschitz functions on compact homogeneous manifolds. Monatsh. Math. **189**(1), 23–49 (2019). ISSN: 0026-9255. https://doi.org/10.1007/s00605-018-1253-0
2. R. Daher, A. Fernandez, J.E. Restrepo, Characterising extended Lipschitz type conditions with moduli of continuity. Results Math. **76**(3), Paper No. 125, 1–18 (2021). ISSN: 1422-6383. https://doi.org/10.1007/s00025-021-01433-2
3. P.L. Duren, *Theory of H^p Spaces*. Pure and Applied Mathematics, vol. 38 (Academic, New York/London, 1970), pp. xii+258
4. B. Fuglede, Commuting self-adjoint partial differential operators and a group theoretic problem. J. Funct. Anal. **16**, 101–121 (1974). https://doi.org/10.1016/0022-1236(74)90072-X
5. A. Hendrickx, L^p-L^q boundedness of Fourier multipliers on fundamental domains of lattices in $\mathbb{R}^d$. J. Fourier Anal. Appl. **28**(60) (2022). https://doi.org/10.1007/s00041-022-09955-1
6. E.C. Titchmarsh, *Introduction to the Theory of Fourier Integrals*, 3rd edn. (Chelsea Publishing Co., New York, 1986), pp. x+394. ISBN: 0-8284-0324-4
7. M.S. Younis, Fourier Transforms of Lipschitz Functions on Compact Groups. Ph.D. Thesis – McMaster University (Canada). ProQuest LLC, Ann Arbor (1974). https://www.proquest.com/docview/302766437
8. M.S. Younis, Fourier transforms of Dini-Lipschitz functions. Int. J. Math. Math. Sci. **9**(2), 301–312 (1986). ISSN: 0161-1712. https://doi.org/10.1155/S0161171286000376
9. A. Zygmund, *Trigonometrical Series* (Dover Publications, New York, 1955), pp. vii+329

Semi-sparsity for Triangular Mesh Denoising

Junqing Huang

Abstract We propose a novel semi-sparsity model for denoising 3D triangular meshes, inspired by the successful application of semi-sparsity regularization in image processing. Firstly, the differential operators defined on the piecewise constant function spaces are introduced for triangular meshes, and the semi-sparsity model is then extended to meshes denoising application. To solve the proposed model efficiently, we also propose an efficient iterative algorithm based on the alternating direction method of multipliers (ADMM) technique. We demonstrate that the semi-sparsity regularization can also be effectively applied to graphic processing tasks, yielding similar simultaneous-fitting outcomes in preserving sharp features and achieving piecewise smoothing surfaces.

1 Introduction

Mesh denoising is a fundamental research topic in the field of geometry processing. Over the years, numerous techniques have been explored to effectively remove noise while preserving the underlying geometric features. These techniques encompass filtering methods, such as bilateral filter [1], guided filter [10], and variational-based approaches [6, 9]. Higher-order variants, such as triangulated surface regularization [4] and advanced mesh-based methods [5], have also been investigated. Unfortunately, these existing methods often introduce staircase artifacts when applied to polynomial-smoothed surfaces. To address this issue, recent advancements in higher-order variational models, such as total generalized variation (TGV), have been proposed to alliterate the artifacts.

In this paper, we propose a higher-order model for denoising 3D triangular meshes. Our approach is based on semi-sparsity regularization, which allows for the

J. Huang (✉)
Department of Mathematics: Analysis, Logic and Discrete Mathematics, Ghent University, Ghent, Belgium
e-mail: Junqing.Huang@UGent.be

M. Ruzhansky, B. Torebek (eds.), *Extended Abstracts MWCAPDE 2023*,
Research Perspectives Ghent Analysis and PDE Center 1,
https://doi.org/10.1007/978-3-031-41665-1_6

preservation of sharp features and the smoothing of surfaces in a piecewise manner. Firstly, we describe the differential operator of piecewise constant function spaces for triangular meshes. Subsequently, we illustrate how the semi-sparsity model can be extended to address the mesh denoising problem. To assess the effectiveness of our proposed method, we introduce an efficient iterative algorithm based on the alternating direction method of multipliers (ADMM) technique. The experimental results demonstrate the superior performance of our proposed method in terms of its ability to preserve local complex graphical features, reinforcing its effectiveness in comparison to existing approaches.

2 Preliminaries

2.1 Notation

Let $\mathcal{M}$ be a non-degenerate triangulated surface with vertices, edges, and triangles denoted as $v_{i(i=0,1,\cdots,I-1)}$, $e_{j(j=0,1,\cdots,E-1)}$ and $\tau_{k(k=0,1,\cdots,T-1)}$, respectively. We introduce the relative orientation of an edge e to a triangle τ, denoted by $s(e,\tau)$, where $v \prec e$ represents that v is an endpoint of an edge e. Similarly, $e \prec \tau$ denotes that e is an edge of a triangle τ; $v \prec \tau$ denotes that v is a vertex of a triangle τ. The relative orientation of an edge e to a triangle τ is denoted by $\operatorname{sgn}(e,\tau)$. We assume that all triangles have the counterclockwise orientations and all edges are with randomly chosen fixed orientations. If an edge e and a triangle τ have the same orientation, then $s(e,\tau) = 1$; otherwise, $s(e,\tau) = -1$. The reader is also referred to [4, 5, 7] for more details.

2.2 Piecewise Linear Function Spaces and Differential Operators

Piecewise constant function spaces have achieved great success in computer graphics applications [4, 5, 9]. We introduce the spaces and show how to derive the associated differential operators on triangulated meshes.

We define the space $U = \mathbb{R}^{\mathrm{T}}$, which is isomorphic to the piecewise constant function space on a triangulated mesh $\mathcal{M}$. Let $u = (u_0, \ldots, u_{\mathrm{T}-1}) \in U$ and u_τ be a vector restricted to triangle τ, sometimes written as $u|_\tau$ for convenience. For example, u_τ can be defined as the outward-facing normal vector restricted on triangle τ, which, as shown in Fig. 1a, is perpendicular to plane defined by triangle τ. According to [4, 5], the jump function of u over an edge e is defined as

$$[u]_e = \begin{cases} \sum_{e \prec \tau} u|_r \operatorname{sgn}(e,\tau), & e \not\subseteq \partial\mathcal{M} \\ 0, & e \subseteq \partial\mathcal{M} \end{cases} \tag{2.1}$$

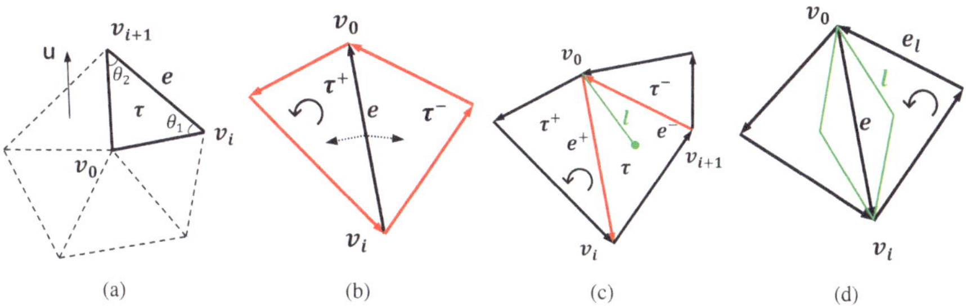

Fig. 1 The illustration of discrete operators on meshes

Here, the jump function $[u]_e$ can be illustrated in Fig. 1b.

The space U has the standard inner product and norm

$$\left(u^1, u^2\right)_U = \sum_\tau u^1\Big|_\tau u^2\Big|_\tau s_\tau, \quad \|u\|_V = \sqrt{(u, u)_U}, \tag{2.2}$$

where u^1, u^2, and $u \in V$ and s_τ is the area of triangle τ.

Let the edge function space $V = \mathbb{R}^E$ and $v_e \in V$ (or $v|_e$) be a vector restricted to edge e; it is then natural to define the first-order differential operator $\mathcal{D} : U \mapsto V$ on $\mathcal{M}$ as

$$\mathcal{D}u|_e = [u]_e, \quad \forall e, \text{ for } u \in V. \tag{2.3}$$

By definition, the space V is also equipped with the inner product and norm

$$\left(v^1, v^2\right)_V = \sum_e v^1\Big|_e v^2\Big|_e \text{len}(e), \quad \|v\|_V = \sqrt{(v, v)_V}, \tag{2.4}$$

where v^1, v^2, $v \in V$, and $len(e)$ is the length of the edge e.

By analogy, it is easy to define the higher-order differential operators. Let l be the line connecting the barycenter and one vertex of the triangle τ. As depicted in Fig. 1c, the two edges e^+ and e^- share the common vertex of l, and the two triangles sharing edges e^+ and e^- are denoted as τ^+ and τ^-, respectively. We then define the jump difference over the line l as $[[u]]_{l,\tau}$ (or, $[[u]]_l$)

$$\begin{aligned}
[[u]]_{l,\tau} &= [u]_{e^+}\, sgn(e^+, \tau^+) + [u]_{e^-}\, sgn(e^-, \tau^-) \\
&= (u_{\tau^+} - u_\tau) - (u_\tau - u_{\tau^-}) \\
&= u_{\tau^+} - 2u_\tau + u_{\tau^-}.
\end{aligned} \tag{2.5}$$

It is clear that Eq. 2.5 can be viewed as a second-order difference operator with respect to u. For any $u \in U$ with the Neumann boundary condition, we actually have

$$[[u]]_l = \begin{cases} u_{\tau+} - 2u_\tau + u_{\tau^-}, & e^+ \text{or } e^- \not\subset \partial\mathcal{M} \\ 0, & e^+ \text{or } e^- \subset \partial\mathcal{M} \end{cases} \tag{2.6}$$

We can see that $[[u]]_l$ is invariant under the choice of orientation of edges. It is also possible to define the higher-order differential operator for the edge e as shown in Fig. 1d

3 Semi-sparsity Regularization for Mesh Denoising

We now extend the semi-sparse model [3] to 3D geometry, which is motivated by the fact that 3D meshes suffer from the similar piecewise constant and smoothing surfaces with discontinuous boundaries.

3.1 Problem Formulation

According to [3], the semi-sparsity prior knowledge of signal is suggested to be formulated into a higher-order L_0 regularization model in the context of optimization-based framework, which has a general following form:

$$\min_u \frac{\beta}{2}\|u - f\|_2^2 + \alpha_1 \sum_{k=1}^{n-1} \left\|\nabla^k u - \nabla^k f\right\|_p^p + \alpha_2 \left\|\nabla^n u\right\|_0 \tag{3.1}$$

where u and f are the target output and observation signals (images, 3D meshes, etc.), respectively. β, α_1, and α_2 weigh the balance of three terms. The idea of Eq. 3.1 is straightforward, that is, a sparse-induced L_0-norm constraint is only imposed on the highest order n-th gradient domain, as the ones with the orders less than n are not fully sparse but also have a small error L_p space.

The above claims are also valid for the piecewise smoothing surfaces of 3D meshes. Given a noisy mesh $\mathcal{M}$ with the normal field denoted as $\mathbf{N}_0$, we have the semi-sparse regularization for normal filter as the following problem:

$$\min_u \frac{\beta}{2}\|\mathbf{N} - \mathbf{N}_0\|_{\mathbf{U}}^2 + \alpha_1 \|\nabla\mathbf{N} - \nabla\mathbf{N}_0\|_{\mathbf{V}} + \alpha_2 \left\|\nabla^2\mathbf{N}\right\|_{\mathbf{W}}^0. \tag{3.2}$$

Here, as indicated in [3], we set highest order $n = 2$ and $p = 1$ in Eq. 3.1 for the sake of simplicity and computational efficiency.

3.2 The Efficient ADMM Solver

Due to the non-smooth and non-convex objective function of Eq.3.2, a direct solution is not available. Instead, we propose to solve the problem based on an alternating direction method of multipliers (ADMM), which has achieved great success in solving the related problems [4, 5].

By introducing the new variables P, and Q, we then reformulate Eq. 3.2 as a constrained optimization problem with the following form:

$$\min_{\mathbf{N},\mathbf{P},\mathbf{Q}} \left\{ \frac{\beta}{2} \|\mathbf{N}-\mathbf{N}_0\|_{\mathbf{U}}^2 + \alpha_1 \|\mathbf{P}\|_{\mathbf{V}} + \alpha_2 \|\mathbf{Q}\|_{\mathbf{W}}^0 + \Psi(\mathbf{N}) \right\}, \quad \text{s.t.} \quad \mathbf{P} = \nabla\mathbf{N} - \nabla\mathbf{N}_0, \quad \mathbf{Q} = \nabla^2\mathbf{N}, \tag{3.3}$$

where

$$\Psi(\mathbf{N}) = \begin{cases} 0, & \|\mathbf{N}_\tau\| = 1, \forall\tau, \\ +\infty, & \text{otherwise}. \end{cases} \tag{3.4}$$

Accordingly, we introduce the augmented Lagrangian function of the above-constrained optimization problem

$$\begin{aligned} \mathcal{L}\left(\mathbf{N},\mathbf{P},\mathbf{Q},\lambda_{\mathbf{P}},\lambda_{\mathbf{Q}}\right) &= \frac{\beta}{2} \|\mathbf{N}-\mathbf{N}_0\|_{\mathbf{U}}^2 + \alpha_1 \|\mathbf{P}\|_{\mathbf{V}} + \alpha_2 \|\mathbf{Q}\|_{\mathbf{W}}^0 + \Psi(\mathbf{N}) \\ &+ \langle \lambda_{\mathbf{P}}, (\nabla\mathbf{N}-\nabla\mathbf{N}_0) - \mathbf{P}\rangle_{\mathbf{V}} + \frac{\rho_1}{2} \|(\mathbf{N}-\mathbf{N}_0) - \mathbf{P}\|_{\mathbf{V}}^2 \\ &+ \langle \lambda_{\mathbf{Q}}, \nabla^2\mathbf{N} - \mathbf{Q}\rangle_{\mathbf{W}} + \frac{\rho_2}{2} \left\|\nabla^2\mathbf{N} - \mathbf{Q}\right\|_{\mathbf{W}}^2, \end{aligned} \tag{3.5}$$

where $\lambda_{\mathbf{P}}$ and $\lambda_{\mathbf{Q}}$ are Lagrange multipliers ρ_1 and ρ_2 are positive penalty weights. The variable-splitting technique is then applied to iteratively update the variables in an alternative way, giving the following subproblems.

3.2.1 The N-Subproblem

$$\begin{aligned} \min_{\mathbf{N}} \frac{\beta}{2} \|\mathbf{N}-\mathbf{N}_0\|_{\mathbf{U}}^2 &+ \frac{\rho_1}{2} \left\|(\nabla\mathbf{N}-\nabla\mathbf{N}_0) - \mathbf{P} + \frac{\lambda_{\mathbf{P}}}{\rho_1}\right\|_{\mathbf{V}}^2 \\ &+ \frac{\rho_2}{2} \left\|\nabla^2\mathbf{N} - \mathbf{Q} + \frac{\lambda_{\mathbf{Q}}}{\rho_2}\right\|_{\mathbf{V}}^2 + \Psi(\mathbf{N}). \end{aligned} \tag{3.6}$$

It is clear that Eq. 3.6 is a quadratic optimization problem with the unit normal constraints. As suggested in [5], an approximation strategy is employed to solve this problem. We first solve the quadratic program and then project the solution **N** onto a unit sphere. Specifically, the corresponding Euler-Lagrange equation based on the first-order optimal conditions has the form

$$\beta\left(\mathbf{N}-\mathbf{N}_0\right)-\rho_1\nabla^*\left(\nabla\mathbf{N}-\nabla\mathbf{N}_0-\mathbf{P}+\frac{\lambda_{\mathbf{P}}}{\rho_1}\right)$$
$$+\rho_2\left(\nabla^2\right)^*\left(\nabla^2\mathbf{N}-\mathbf{Q}+\frac{\lambda_{\mathbf{Q}}}{\rho_2}\right)=0. \tag{3.7}$$

where ∇^* and $\left(\nabla^2\right)^*$ are the adjoint operators of the first-order and second-order differential operators, respectively. The above equation is a sparse and positive semi-definite linear system, which can be solved by efficient sparse linear solvers.

3.2.2 The P-Subproblem

$$\min_{\mathbf{P}}\alpha_1\left\|\mathbf{P}\right\|_{\mathbf{V}}+\frac{\rho_1}{2}\left\|\mathbf{P}-\left(\nabla\mathbf{N}+\nabla\mathbf{N}_0\right)+\frac{\lambda_{\mathbf{P}}}{\rho_1}\right\|_{\mathbf{V}}^2, \tag{3.8}$$

where (3.8) is a classical Lasso problem and can be solved efficiently by splitting each variable $\mathbf{P}_e$ independently, where the $\mathbf{P}_e$ has a closed form solution

$$\mathbf{P}_e=\mathcal{S}\left(\left(\nabla\mathbf{N}+\nabla\mathbf{N}_0\right)-\frac{\lambda_{\mathbf{P}}}{\rho_1},\frac{\alpha_1}{\rho_1}\right), \tag{3.9}$$

with the soft shrinkage operator $\mathcal{S}(x,T)$ defined as

$$\mathcal{S}(x,T)=\mathrm{sign}(x)\max\left(0,\|x\|-T\right).$$

3.2.3 The Q-Subproblem

$$\min_{\mathbf{Q}}\alpha_2\left\|\mathbf{Q}\right\|_{\mathbf{W}}^0+\frac{\rho_2}{2}\left\|\mathbf{Q}-\nabla^2\mathbf{N}+\frac{\lambda_{\mathbf{Q}}}{\rho_2}\right\|_{\mathbf{V}}^2. \tag{3.10}$$

The **Q**-subproblem (3.10) is a L_0 norm minimization problem, which has a similar separable property as (3.8), and each variable $\bar{Q}_l$ is given by the formula

$$\mathbf{Q}_l=\mathcal{H}\left(\nabla^2\mathbf{N}-\frac{\lambda_{\mathbf{Q}}}{\rho_2},\frac{\alpha_2}{\rho_2}\right), \tag{3.11}$$

with the hard-threshold operator defined as

$$\mathcal{H}(x, T) = \begin{cases} 0, & \|x\| \le T, \\ x, & \text{otherwise.} \end{cases}$$

Finally, the Lagrange multipliers $\lambda_{\mathbf{P}}$ and $\lambda_{\mathbf{Q}}$ are updated in the form

$$\begin{aligned} \lambda_{\mathbf{P}} &= \lambda_{\mathbf{P}} + \rho_1 \left((\nabla \mathbf{N} - \nabla \mathbf{N}_0) - \mathbf{P} \right), \\ \lambda_{\mathbf{Q}} &= \lambda_{\mathbf{Q}} + \rho_2 \left(\nabla^2 \mathbf{N} - \mathbf{Q} \right). \end{aligned} \tag{3.12}$$

In summary, the semi-sparsity model in Eq. 3.2 for normal filter is achieved by solving the ADMM subproblems and Lagrange multipliers iteratively. The procedure terminates when one of the stopping criteria is met.

4 Experimental Results

We have interpreted the definitions of the differential operators on triangular meshes in the previous section. Once they are computed, it is easy to substitute them into the semi-sparsity model and solve it based on the ADMM algorithm accordingly. In order to further illustrate the proposed semi-sparsity model, we here compare it with the existing mesh denoising methods, including the feature-aware mesh filter [8], bilateral filter (BF) [1], L_0 minimization [2], and high-order TGV regularization [5].

As shown in Fig. 2, the original surface contains corners, edges, and polynomial smoothing surfaces. The BF method removes noise in smoothing areas but also seriously blurs sharp features; and L_0 minimization retains sharp features but leads to slanted artifacts in the smoothing regions. However, our method not only produces a similar result as the cutting-edge high-order (HO) regularization [4] in polynomial-smoothing regions but also preserves the sharpening corners and edges. This is further demonstrated by the results in Fig. 3. The experiments demonstrate the extension of our semi-sparse model to triangular meshes. We

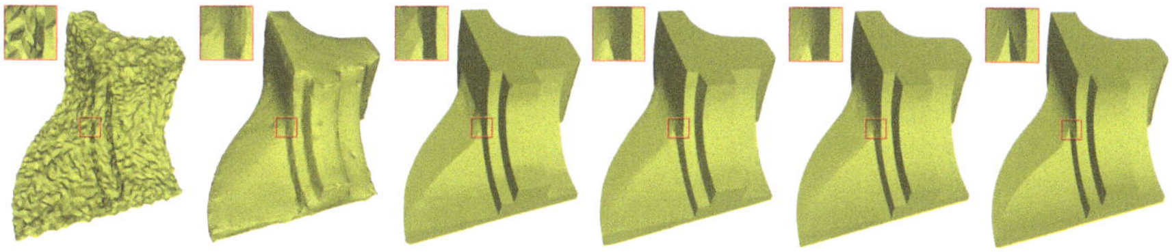

(a) Noisy (b) BF[1] (c) L_0[2] (d) TGV[5] (e) Ours (f) GT

Fig. 2 Visual comparison of mesh denoising. (**a**) Input (Gaussian noise $\sigma = 0.3\bar{L}$), (**b**) Bilateral filter (BF) [1], (**c**) L_0 minimization [2], (**d**) TGV regularization [4], (**e**) Our result, and (**f**) Ground truth (GT)

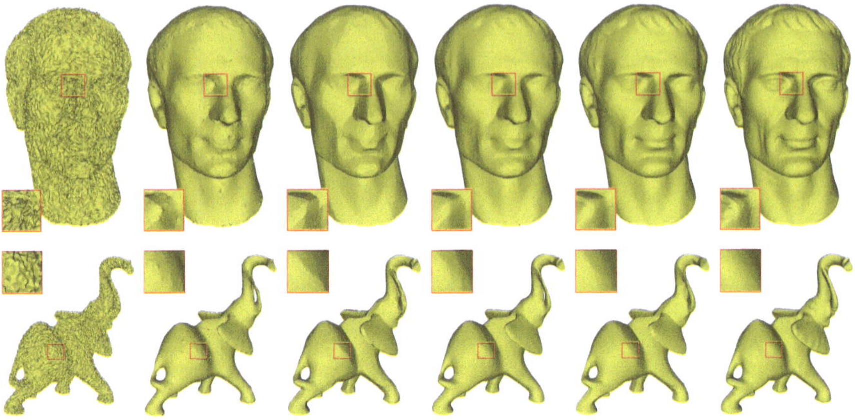

Fig. 3 Mesh denoising results. (**a**) Input (Gaussian noise $\sigma = 0.3\bar{L}$), (**b**) Bilateral filter (BF) [1], (**c**) L_0 minimization [2], (**d**) TGV regularization [5], (**e**) Our result, and (**f**) Ground truth (GT)

define the differential operator over the edge and only update the vertexes for mesh denoising. We refer the interested reader to [4, 5, 7] for more details.

Acknowledgments The research is financially supported by the Research Foundation – Flanders (FWO) Odysseus 1 under Grant G.0H94.18N; Methusalem Programme of the Ghent University Special Research Fund (BOF) under Grant 01M01021.

References

1. S. Fleishman, I. Drori, D. Cohen-Or, Bilateral mesh denoising, in *ACM Transactions on Graphics (TOG)*, vol. 22 (ACM, 2003), pp. 950–953
2. L. He, S. Schaefer, Mesh denoising via L_0 minimization. ACM Trans. Graph. (TOG) **32**(4), 64 (2013)
3. J. Huang, H. Wang, X. Wang, M. Ruzhansky, Semi-sparsity for smoothing filters. IEEE Trans. Image Process. **32**, 1627–1639 (2023)
4. Z. Liu, R. Lai, H. Zhang, C. Wu, Triangulated surface denoising using high order regularization with dynamic weights. SIAM J. Sci. Comput. **41**(1), B1–B26 (2019)
5. Z. Liu, Y. Li, W. Wang, L. Liu, R. Chen, Mesh total generalized variation for denoising. IEEE Trans. Vis. Comput. Graph. **28**(12), 4418–4433 (2021)
6. Z. Liu, W. Wang, S. Zhong, B. Zeng, J. Liu, W. Wang, Mesh denoising via a novel Mumford-Shah framework. Comput.-Aided Des. **126**, 102858 (2020)
7. M. Meyer, M. Desbrun, P. Schröder, A.H. Barr, Discrete differential-geometry operators for triangulated 2-manifolds, in *Visualization and Mathematics III* (Springer, Berlin, 2003), pp. 35–57.
8. X. Sun, P.L. Rosin, R. Martin, F. Langbein, Fast and effective feature-preserving mesh denoising. IEEE Trans. Vis. Comput. Graph. **13**(5), 925–938 (2007)

9. H. Zhang, C. Wu, J. Zhang, J. Deng, Variational mesh denoising using total variation and piecewise constant function space. IEEE Trans. Vis. Comput. Graph. **21**(7), 873–886 (2015)
10. W. Zhang, B. Deng, J. Zhang, S. Bouaziz, L. Liu, Guided mesh normal filtering, in *Computer Graphics Forum*, vol. 34 (Wiley Online Library, 2015), pp. 23–34

Super-Resolution Reconstruction from Truncated Fourier Transform

Mikhail Isaev, Roman G. Novikov, and Grigory V. Sabinin

Abstract We present recent theoretical and numerical results on recovering a compactly supported function v on $\mathbb{R}^d$, $d \geq 1$, from its Fourier transform $\mathcal{F}v$ given within the ball B_r. We proceed from known results on the prolate spheroidal wave functions and on the Radon transform. The most interesting point of our numerical examples consists in super-resolution, that is, in recovering details beyond the diffraction limit, that is, details of size less than π/r, where r is the radius of the ball mentioned above. This short review is based on the works Isaev, Novikov (2022 J. Math. Pures Appl. 163 318–333) and Isaev, Novikov, Sabinin (2022 Inverse Problems 38 105002).

2000 Mathematics Subject Classification 42A38, 35R30, 49K40

1 Introduction

It is well-known that there is a diffraction limit for recovering the structure of an object from results of its probing by waves of length λ. This limit is that the details of the object smaller than $\lambda/2$ are blurred or not recovered by standard numerical methods. Our recent results of [1, 2] give a mathematical approach to overcoming this limit, thereby achieving *super-resolution*, without significant a priori assumptions on the unknown structure. In particular, the details in the

M. Isaev
School of Mathematics, Monash University, Clayton, VIC, Australia
e-mail: mikhail.isaev@monash.edu

R. G. Novikov (✉)
CMAP, CNRS, Ecole polytechnique, Institut polytechnique de Paris, Palaiseau, France
IEPT RAS, Moscow, Russia
e-mail: novikov@cmap.polytechnique.fr

G. V. Sabinin
Faculty of Mechanics and Mathematics, Lomonosov Moscow State University, Moscow, Russia

M. Ruzhansky, B. Torebek (eds.), *Extended Abstracts MWCAPDE 2023*,
Research Perspectives Ghent Analysis and PDE Center 1,
https://doi.org/10.1007/978-3-031-41665-1_7

numerical examples of [2] are distinguished even of the size $\lambda/12$ (i.e. 6 times smaller than the diffraction limit) and even with 20% random noise in the data.

In [1, 2], the original problem is modelled as a problem of recovering a function with compact support on $\mathbb{R}^d$ from its Fourier transform restricted to the ball B_r of fixed radius $r = 2\pi/\lambda$ (i.e., from the truncated Fourier transform); see Problem 1. This model arises, for example, as a reconstruction problem in monochromatic acoustic tomography in the Born approximation.

Thus, we consider the Fourier transform $\mathcal{F}$ defined by the formula

$$\mathcal{F}[v](p) = \hat{v}(p) := \frac{1}{(2\pi)^d} \int\limits_{\mathbb{R}^d} e^{ipq} v(q) dq, \qquad p \in \mathbb{R}^d, \tag{1.1}$$

where v is a complex-valued test function on $\mathbb{R}^d$, $d \geq 1$.

Let $B_\rho := \left\{q \in \mathbb{R}^d : |q| < \rho\right\}$, $\rho > 0$.

We now state formally the aforementioned problem.

Problem 1 Find $v \in L^2(\mathbb{R}^d)$, where *supp* $v \subset B_\sigma$, from $\hat{v} = \mathcal{F}v$ given on the ball B_r (possibly with some noise), for fixed $r, \sigma > 0$.

Problem 1 arises in different areas and has been extensively studied in the literature. However, solving Problem 1 is complicated considerably by the fact it is exponentially unstable, for fixed $r, \sigma > 0$. Nevertheless, there exist several techniques to approach this problem theoretically and numerically; see, [1, 2] and references therein.

The conventional approach for solving Problem 1 is based on the following approximation:

$$v \approx v_{\text{naive}} := \mathcal{F}^{-1}[w](q) = \int\limits_{B_r} e^{-ipq} w(p) dp \qquad q \in B_\sigma, \tag{1.2}$$

where $\mathcal{F}^{-1}$ is the standard inverse Fourier transform and w is such that $w|_{B_r}$ coincides with the data of Problem 1 and $w|_{\mathbb{R}^d \setminus B_r} \equiv 0$.

Formula (1.2) leads to a stable and accurate reconstruction for sufficiently large r, which is carefully studied in the literature and widely used for industrial applications. Nevertheless, it has the well-known diffraction limit: small details (especially less than π/r) are blurred.

A new approach for *super-resolution* in comparison with the resolution of (1.2) was recently developed in [1, 2]; see Sects. 3 and 4. Some preliminaries are presented in Sect. 2.

2 Preliminaries

For convenience, we consider the scaling of v with respect to the size of its support:

$$v_\sigma(q) := v(\sigma q), \qquad q \in \mathbb{R}^d. \tag{2.1}$$

Note that $supp\ v_\sigma \subset B_1$. Let

$$c := r\sigma. \tag{2.2}$$

The data in Problem 1 (for the case without noise) can be presented as follows (for more details, see [1]):

$$\hat{v}(rx) = \frac{\sigma}{2\pi}\mathcal{F}_c\left[v_\sigma\right](x) \qquad \text{for } d = 1, \tag{2.3}$$

$$\hat{v}(rx\theta) = \left(\frac{\sigma}{2\pi}\right)^d \mathcal{F}_c\left[\mathcal{R}_\theta[v_\sigma]\right](x) \qquad \text{for } d \geq 2, \tag{2.4}$$

where $x \in [-1, 1]$, $\theta \in \mathbb{S}^{d-1}$, and the operators $\mathcal{F}_c$ and $\mathcal{R}_\theta$ are defined by

$$\mathcal{F}_c[f](x) := \int_{-1}^{1} e^{icxy} f(y)dy, \qquad x \in [-1, 1], \tag{2.5}$$

$$\mathcal{R}_\theta[u](y) := \int\limits_{q\in\mathbb{R}^d, q\theta=y} u(q)dq, \qquad y \in \mathbb{R}, \tag{2.6}$$

where f is a test function on $[-1, 1]$ and u is a test function of $\mathbb{R}^d$.

Recall that $\mathcal{R}_\theta[u] \equiv \mathcal{R}[u](\cdot, \theta)$, where $\mathcal{R}_\theta$ is defined by (2.6) and $\mathcal{R}$ is the classical *Radon transform* of [4]. In fact, presentation (2.4) follows from the projection theorem of the Radon transform theory.

The operator $\mathcal{F}_c$ defined by (2.5) is a variant of band-limited Fourier transform. This operator is one of the key objects of the theory of *prolate spheroidal wave functions*. In particular, the operator $\mathcal{F}_c$ has the following singular value decomposition in $L^2([-1, 1])$:

$$\mathcal{F}_c[f](x) = \sum_{j\in\mathbb{N}} \mu_{j,c}\psi_{j,c}(x) \int_{-1}^{1} \psi_{j,c}(y) f(y)dy, \tag{2.7}$$

where $(\psi_{j,c})_{j\in\mathbb{N}}$ are the prolate spheroidal wave functions (these functions reduce to the Legendre polynomials as $c \to 0$ and reduce to the Hermite functions as $c \to \infty$). In addition, it is known that the eigenvalues $\{\mu_{j,c}\}_{j\in\mathbb{N}}$ satisfy the following properties:

$$0 < |\mu_{j+1,c}| < |\mu_{j,c}| \text{ for all } j \in \mathbb{N} = \{0, 1, 2 \ldots\}; \tag{2.8}$$

$$\left\lfloor \frac{2c}{\pi} \right\rfloor - 1 \leq \left|\{j \in \mathbb{N},\ |\mu_{j,c}| \geq \sqrt{\pi/c}\}\right| \leq \left\lceil \frac{2c}{\pi} \right\rceil + 1, \tag{2.9}$$

where $\lfloor \cdot \rfloor$ and $\lceil \cdot \rceil$ denote the floor and the ceiling functions, respectively, and $|\cdot|$ in (2.9) is the number of elements in a set;

$$\mu_{j,c} \text{ decay superexponentially as } j \to \infty. \tag{2.10}$$

The functions $(\psi_{j,c})_{j\in\mathbb{N}}$ are certain of wave functions introduced by Niven in [3] for solving the Helmholtz equation in prolate spheroidal coordinates. Originally, $(\psi_{j,c})_{j\in\mathbb{N}}$ are defined as the eigenfunctions of the spectral problem

$$\frac{d}{dx}\left[(1-x^2)\frac{d\psi}{dx}\right] + c^2x^2\psi = \chi\psi, \qquad \psi \in C^2([-1,1]). \tag{2.11}$$

The fact that $(\psi_{j,c})_{j\in\mathbb{N}}$ are the eigenfunctions of the finite Fourier transform $\mathcal{F}_c$ defined by (2.5) was pointed out by Slepian and Pollak in [5] as a special case of more general integral relations satisfied by Niven's wave functions. As mentioned in [5]:

> *These functions ... possess properties that make them ideally suited for the study of certain questions regarding the relationship between functions and their Fourier transforms.*

Surprisingly, to our knowledge, these functions were not used in the context of Problem 1 before the recent work [1], at least, for $d > 1$.

3 Reconstruction Formulas from [1]

The formulas of [1] for solving Problem 1 are as follows, where we use the notations of Sect. 2.

For $d = 1$, the reconstruction is given by

$$v_\sigma = \frac{2\pi}{\sigma}\mathcal{F}_c^{-1}[\hat{v}_r], \tag{3.1}$$

where $\hat{v}_r(x) = \hat{v}(rx)$, $x \in [-1,1]$, and

$$\mathcal{F}_c^{-1}[g](y) = \sum_{j\in\mathbb{N}} \frac{1}{\mu_{j,c}}\psi_{j,c}(y)\int_{-1}^{1}\psi_{j,c}(x)g(x)dx, \tag{3.2}$$

where g is a test function from the range of $\mathcal{F}_c$ acting on $L^2([-1,1])$.

For $d \geq 2$, the reconstruction is given by

$$v_\sigma = \left(\frac{2\pi}{\sigma}\right)^d \mathcal{R}^{-1}[f_{r,\sigma}], \tag{3.3}$$

where

$$f_{r,\sigma}(y,\theta) = \begin{cases} \mathcal{F}_c^{-1}[\hat{v}_{r,\theta}](y), & \text{if } y \in [-1,1], \\ 0, & \text{otherwise;} \end{cases}$$

$$\hat{v}_{r,\theta}(x) = \hat{v}(rx\theta), \qquad x \in [-1,1],\ \theta \in \mathbb{S}^{d-1}.$$

In addition, in view of property (2.10), for the case of noisy data in Problem 1, the operator $\mathcal{F}_c^{-1}$ is approximated by the finite rank operator $\mathcal{F}_{n,c}^{-1}$ defined by

$$\mathcal{F}_{n,c}^{-1}[g](y) := \sum_{j=0}^{n} \frac{1}{\mu_{j,c}} \psi_{j,c}(y) \int_{-1}^{1} \psi_{j,c}(x) g(x) dx. \tag{3.4}$$

It is important to remark that:

- the operator $\mathcal{F}_{n,c}^{-1}$ is correctly defined on $L^2([-1,1])$ for any $n \in \mathbb{N}$;
- the function $\mathcal{F}_{n,c}^{-1}[g]$ is *the quasi-solution* in the sense of Ivanov of the equation $\mathcal{F}_c[f] = g \in L^2([-1,1])$ on the span of the first $n+1$ functions $(\psi_{j,c})_{j \le n}$;
- the rank n is a regularisation parameter and its choice is crucial for both theoretical results and numerical applications.

For formulations of mathematical theorems related with formulas (3.1), (3.2), (3.3), and (3.4), see [1]. These theoretical results include Hölder-logarithmic stability estimates for the reconstruction based on these formulas.

4 Numerical Results of [2]

The numerical implementation of [2] of the approach of [1] includes, in particular, different principles for choosing the aforementioned regularisation parameter n such as *residual minimisation* and the *Morozov discrepancy principle*. One of the most interesting points of the results of [2] lies in examples of super-resolution, that is, recovering details of size less than π/r, where r is the band-limiting radius of Problem 1. In [2] we also obtain a better reconstruction in the sense of L^2-norm than the conventional reconstruction based on formula (1.2).

We illustrate numerical results of [2] by Figs. 1 and 2. In these numerical examples, we use the values $\sigma = 1$ and $c = r = 10$; see [2] for a detailed description of the reconstructions.

Figure 1 shows one of the numerical examples of [2] for $d = 1$. In this example, the preimage v equals the sum of characteristic functions of two disjoint intervals at distance $\pi/(2r)$, which is twice smaller than the diffraction limit. Figure 1 shows our super-resolution reconstruction based on formulas (3.1), (3.4) in comparison

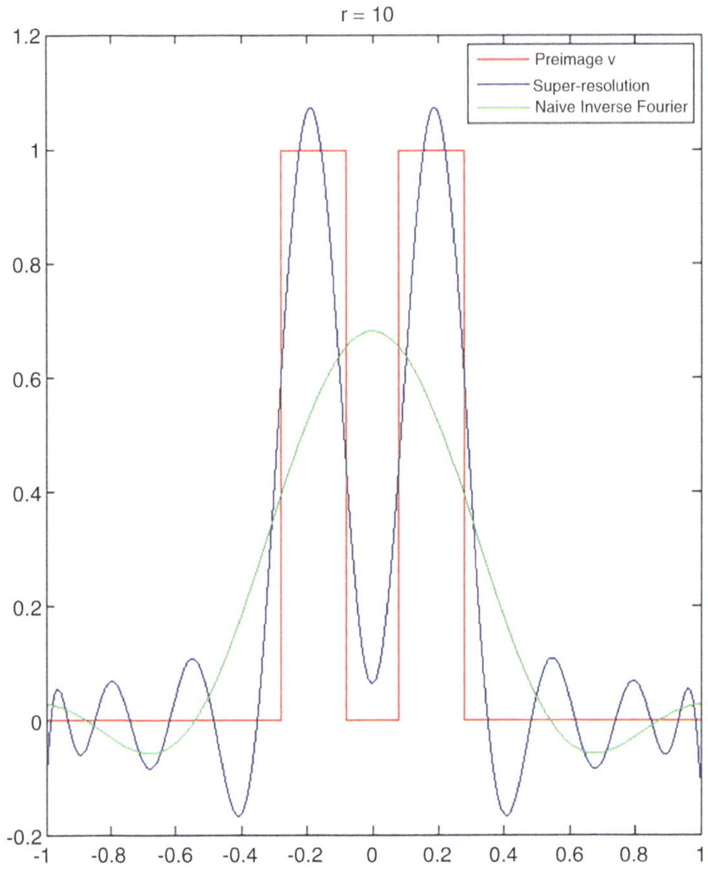

Fig. 1 Example of our super-resolution reconstruction in comparison with preimage v and naive Fourier inversion, for $d = 1$

with preimage v and naive Fourier inversion v_{naive} based on formula (1.2), where the data are noiseless.

Figure 2 shows one of the numerical examples of [2] for $d = 2$. In this example, the preimage v equals the sum of characteristic functions of three disjoint square parts. The distances between the parts of v are significantly less than $\pi/r \approx 0.314$: two bottom squares are at the distance 0.1, while the top square and any of the bottom squares are at the distance 0.05, which is 6 times smaller than the diffraction limit. Figure 2 shows our super-resolution reconstruction based on formulas (3.3)–(3.4) in comparison with preimage v and naive Fourier inversion v_{naive} based on formula (1.2), where the data are noiseless.

Our work [2] also contains many other numerical results and examples. In particular, in spite of the exponential instability of Problem 1, we achieve super-resolution even for noisy data by an appropriate choice of the regularisation

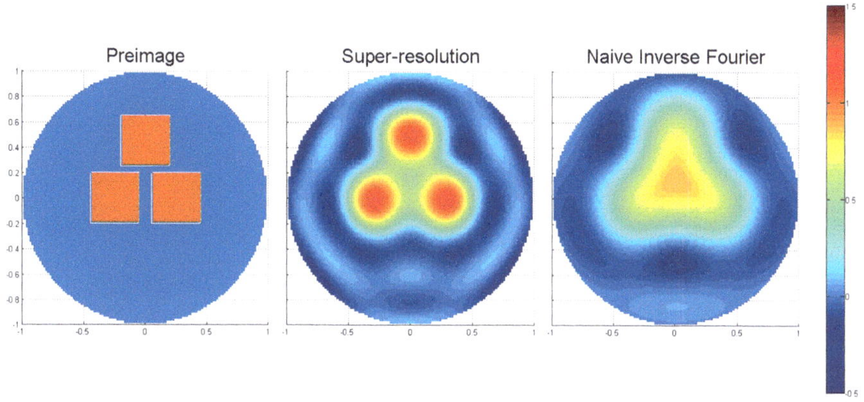

Fig. 2 Example of our super-resolution reconstruction in comparison with preimage v and naive Fourier inversion, for $d = 2$

parameter n. In fact, our reconstruction with $n = n_0 = \left\lfloor \frac{2c}{\pi} \right\rfloor$ behaves similarly to the conventional reconstruction based on formula (1.2). Taking n properly larger than n_0 leads to better results. Moreover, for $d \geq 2$, our approach works well even for a considerable level of random noise: for example, the reconstruction given in Fig. 2 remains reliable even if we add 20% random noise to the data.

Acknowledgments The research is partially supported by ARC DE200101045.

References

1. M. Isaev, R.G. Novikov, Reconstruction from the Fourier transform on the ball via prolate spheroidal wave functions. J. Math. Pures et Appl. **163**, 318–333 (2022)
2. M. Isaev, R.G. Novikov, G.V. Sabinin, Numerical reconstruction from the Fourier transform on the ball using prolate spheroidal wave functions. Inverse Problems **38**, 105002 (2022)
3. C. Niven, On the conduction of heat in ellipsoids of revolution. Philos. Trans. R. Soc. Lond. **171**, 117–151 (1880)
4. J. Radon, Uber die Bestimmung von Funktionen durch ihre Integralwerte längs gewisser Mannigfaltigkeiten. Ber. Saechs Akad. Wiss. Leipzig, Math-Phys. **69**, 262–267 (1917)
5. D. Slepian, H.O. Pollak, Prolate spheroidal wave functions, Fourier analysis and uncertainty I. Bell System Tech. J. **40**(1), 43–63 (1961)

Interpolation Theorem for Discrete Net Spaces

Aitolkyn H. Kalidolday and Erlan D. Nursultanov

Abstract In this paper, we study the interpolation properties of the net spaces $n_{p,q}(M)$, when M is the set of all segments from $\mathbb{Z}$. It is shown that in this case, the scale of spaces is closed with respect to the real interpolation method. As a corollary, an interpolation theorem of Marcinkevich type is presented.

AMS Mathematics Subject Classification: 46B70.

1 Let S be the set of all finite sets of indices from $\mathbb{Z}^n$. For a fixed set $M \subset S$, we define the space $n_{p,q}(M)$ $(0 < p, q \leqslant \infty)$ as the set of sequences $a = \{a_m\}_{m\in\mathbb{Z}^n}$ with quasinorm for $0 < p < \infty$, $0 < q < \infty$

$$\|a\|_{n_{p,q}(M)} = \left(\sum_{k=1}^{\infty} k^{\frac{q}{p}-1}(\bar{a}_k(M))^q\right)^{\frac{1}{q}},$$

and for $q = \infty$, $0 < p \leqslant \infty$

$$\|a\|_{n_{p,\infty}(M)} = \sup_{1\leqslant k<\infty} k^{\frac{1}{p}}\bar{a}_k(M),$$

where

$$\bar{a}_k(M) = \sup_{\substack{e\in M\\|e|\geqslant k}} \frac{1}{|e|}\left|\sum_{m\in e} a_m\right|,$$

A. H. Kalidolday
Institute of Mathematics and Mathematical Modeling, Almaty, Kazakhstan

E. D. Nursultanov (✉)
Department of Mechanics and Mathematics, M.V.Lomonosov Moscow State University (Kazakhstan branch), Astana, Kazakhstan

Institute of Mathematics and Mathematical Modeling, Almaty, Kazakhstan
e-mail: er-nurs@yandex.kz

M. Ruzhansky, B. Torebek (eds.), *Extended Abstracts MWCAPDE 2023*, Research Perspectives Ghent Analysis and PDE Center 1,
https://doi.org/10.1007/978-3-031-41665-1_8

where $|e|$ is the number of indices in e.

These spaces were introduced in [10], and they were called net spaces.

Net spaces have found important applications in various problems of harmonic analysis, operator theory, and theory of stochastic processes [1–3, 11–15]. In this paper, we study the interpolation properties of these spaces. It should be noted here that net spaces are in a sense close to the discrete Morrey spaces:

$$m_p^\lambda = \left\{ a = \{a_k\}_{k\in\mathbb{Z}} : \quad \sup_{m\in\mathbb{N}} \sup_{k\in\mathbb{Z}} \frac{1}{m^\lambda} \left(\sum_{r=k}^{k+m} |a_r|^p \right)^{\frac{1}{p}} < \infty \right\}.$$

In the case when $a = \{a_k\}_{k\in\mathbb{Z}}$, $a_k \geqslant 0$, for $\lambda = n\left(1 - \frac{1}{p}\right)$

$$\|a\|_{n_{p,\infty}(M)} \asymp \|a\|_{m_1^\lambda}.$$

The question of interpolation of Morrey spaces was considered in the works [9, 16], and it was shown that this scale of spaces is not closed with respect to the real interpolation method.

In this paper, we show that if M is the set of all segments from $\mathbb{Z}$, the scale of spaces is closed with respect to the real interpolation method, i.e., the following relation holds:

$$\left(n_{p_0,q_0}(M), n_{p_1,q_1}(M)\right)_{\theta,q} = n_{p,q}(M). \tag{1}$$

Here, we also used ideas from [5–8].

Given functions F and G, in this paper $F \lesssim G$ means that $F \leqslant c\, G$ (or $c\, F \geqslant G$), where c is a positive number, depending only on numerical parameters, that may be different on different occasions. Moreover, $F \asymp G$ means that $F \lesssim G$ and $G \lesssim F$.

2 Let (A_0, A_1) be a compatible pair of Banach spaces [4]. Let

$$K(t, a; A_0, A_1) = K(t, a) = \inf_{a=a_0+a_1} (\|a_0\|_{A_0} + t\|a_1\|_{A_1}), \ a \in A_0 + A_1,$$

be the functional Petre. For $0 < q < \infty$, $0 < \theta < 1$

$$(A_0, A_1)_{\theta,q} = \left\{ a \in A_0 {+} A_1 : \|a\|_{(A_0,A_1)_{\theta,q}} = \left(\int_0^\infty (t^{-\theta} K(t, a))^q \frac{dt}{t} \right)^{1/q} < \infty \right\},$$

and for $q = \infty$

$$(A_0, A_1)_{\theta,q} = \left\{ a \in A_0 + A_1 : \|a\|_{(A_0,A_1)_{\theta,q}} = \sup_{0<t<\infty} t^{-\theta} K(t, a) < \infty \right\}.$$

Theorem 1 *Let $1 \leqslant p_0 < p_1 < \infty$ and $0 < q_0, q_1, q \leqslant \infty$. Let M be the set of all segments from $\mathbb{Z}$. Then*

$$\left(n_{p_0,q_0}(M), n_{p_1,q_1}(M)\right)_{\theta,q} = n_{p,q}(M),$$

where $\frac{1}{p} = \frac{1-\theta}{p_0} + \frac{\theta}{p_1}$, $\theta \in (0,1)$.

Proof Let us prove first

$$\left(n_{1,\infty}(M), n_{\infty,\infty}(M)\right)_{\theta,q} = n_{p,q}(M), \tag{2}$$

where $\frac{1}{p} = 1-\theta$, $\theta \in (0,1)$. Let $\tau \in \mathbb{N}$ and M be the set of all segments from $\mathbb{Z}$. Divide our entire axis into disjoint segments $\{I_k\}_{k\in\mathbb{Z}}$, where $I_k = (2^k, 2^{k+1})$ of the measure $|I_k| = \tau$. It is obvious that $\tau = 2^{k+1} - 2^k = 2^k$. Let $a = \{a_m\}_{m\in\mathbb{Z}} \in n_{p,q}(M)$, define the sequence $b_0 = \{b_n^0\}$:

$$b_n^0 = \frac{1}{|I_k|}\left|\sum_{m\in I_k} a_m\right| \quad \text{for } n \in I_k.$$

Note that $|b_n^0| \leqslant \bar{a}_\tau(M)$ and

$$\sum_{n\in I_k}(a_n - b_n^0) = 0.$$

Let $D_\tau(a) = \{b = \{b_n\}_{n\in\mathbb{Z}} : |b_n| \leqslant \bar{a}_\tau(M)\}$. For the Petre functional, we have the following:

$$\begin{aligned}
K(t, a; n_{1,\infty}(M), n_{\infty,\infty}(M)) &= \inf_{a=a_0+a_1}\left(\|a_0\|_{n_{1,\infty}(M)} + t\|a_1\|_{n_{\infty,\infty}(M)}\right)\\
&\leqslant \inf_{b\in D_\tau(a)}\Big(\sup_{1\leqslant k<\infty} k\overline{(a-b)}_k(M) + t\sup_{1\leqslant k<\infty}\bar{b}_k^0(M)\Big)\\
&\leqslant \sup_{1\leqslant k<\infty} k\overline{(a-b^0)}_k(M) + t\bar{a}_\tau(M).
\end{aligned}$$

Consider the first term

$$\sup_{1\leqslant k<\infty} k\overline{(a-b^0)}_k(M) \asymp \sup_{1\leqslant k<\tau} k\overline{(a-b^0)}_k(M) + \sup_{\tau\leqslant k<\infty} k\overline{(a-b^0)}_k(M).$$

Let I be an arbitrary segment from M such that $|I| \geqslant \tau$. Hence

$$\left|\sum_{n\in I}(a_n - b_n^0)\right| = \left|\sum_{I_k\subset I}\sum_{n\in I_k}(a_n - b_n^0) + \sum_{n\in I\cap I_{k_0}}(a_n - b_n^0) + \sum_{n\in I\cap I_{k_1}}(a_n - b_n^0)\right|$$

$$\leqslant \left|\sum_{n\in I\cap I_{k_0}} a_n\right| + \left|\sum_{n\in I\cap I_{k_1}} a_n\right| + \left(|I\cap I_{k_0}| + |I\cap I_{k_1}|\right)\bar{a}_\tau(M),$$

i.e., the following estimate holds:

$$\left|\sum_{n\in I}(a_n - b_n^0)\right| \leqslant s_1\bar{a}_{s_1} + s_2\bar{a}_{s_2} + 2\tau\bar{a}_\tau(M),$$

where $s_1 = |I\cap I_{k_0}| \leqslant \tau$, $s_2 = |I\cap I_{k_1}| \leqslant \tau$. Hence we have

$$\left|\sum_{n\in I}(a_n - b_n^0)\right| \leqslant 4 \sup_{\tau\geqslant s>1} s\,\bar{a}_s(M).$$

Hence,

$$\sup_{\tau\leqslant k<\infty} k\overline{(a-b^0)}_k = \sup_{\tau\leqslant k<\infty} k \sup_{|I|\geqslant k}\frac{1}{|I|}\left|\sum_{n\in\mathbb{Z}}(a_n - b_n^0)\right| \leqslant 4\sup_{\tau\geqslant s\geqslant 1} s\bar{a}_s(M).$$

For the first term, we have

$$\sup_{1\leqslant k<\infty} k\overline{(a-b^0)}_k \leqslant c\sup_{\tau\geqslant k\geqslant 1} k\bar{a}_k(M).$$

Thus,

$$K(t,a;n_{1,\infty}(M),n_{\infty,\infty}(M)) \lesssim \sup_{\tau\geqslant k>1} k\bar{a}_k(M) + t\bar{a}_\tau(M).$$

Then we have

$$\|a\|_{(n_{1,\infty}(M),n_{\infty,\infty}(M))_{\theta,q}} \asymp \left(\sum_{m\in\mathbb{Z}}\left(2^{-\theta m}K(2^m,a)\right)^q\right)^{\frac{1}{q}}$$

$$= \left(\sum_{m=-\infty}^{-1}\left(2^{-\theta m}K(2^m,a)\right)^q\right)^{\frac{1}{q}} + \left(\sum_{m=0}^{\infty}\left(2^{-\theta m}K(2^m,a)\right)^q\right)^{\frac{1}{q}}.$$

For the first term, we have

$$\left(\sum_{m=-\infty}^{-1}\left(2^{-\theta m}K(2^m,a)\right)^q\right)^{\frac{1}{q}}=\left(\sum_{m=-\infty}^{-1}\left(2^{-\theta m}\inf_{a=a_0+a_1}(\|a_0\|_{n_{1,\infty}(M)}\right.\right.$$

$$\left.\left.+2^m\|a_1\|_{n_{\infty,\infty}(M)})\right)^q\right)^{\frac{1}{q}}$$

$$\leqslant\left(\sum_{m=-\infty}^{-1}\left(2^{(1-\theta)m}\|a\|_{n_{\infty,\infty}(M)}\right)^q\right)^{\frac{1}{q}}=c\|a\|_{n_{\infty,\infty}(M)}\lesssim\|a\|_{n_{p,q}(M)}.$$

For the second term, taking into account that $\tau=2^m$ and applying the above estimate for the Petre functional and Minkowski's inequality, we obtain

$$\left(\sum_{m=0}^{\infty}\left(2^{-\theta m}K(2^m,a)\right)^q\right)^{\frac{1}{q}}\leqslant\left(\sum_{m=0}^{\infty}\left(2^{-\theta m}\sup_{\tau\geqslant k>1}k\bar{a}_k(M)+2^m\bar{a}_\tau(M)\right)^q\right)^{\frac{1}{q}}$$

$$\leqslant\left(\sum_{m=0}^{\infty}\left(2^{-\theta m}\sup_{\tau\geqslant k>1}k\bar{a}_k(M)\right)^q\right)^{\frac{1}{q}}+\left(\sum_{m=0}^{\infty}\left(2^{(1-\theta)m}\bar{a}_{2^m}(M)\right)^q\right)^{\frac{1}{q}}.$$

Considering that $k\asymp\left(\sum_{r=1}^{2^m}r^{q-1}\right)^{\frac{1}{q}}$ and changing the order of summation, we get

$$\left(\sum_{m=0}^{\infty}\left(2^{-\theta m}K(2^m,a)\right)^q\right)^{\frac{1}{q}}\leqslant\left(\sum_{m=0}^{\infty}\left(2^{-\theta m}\left(\sum_{r=1}^{2^m}r^{q-1}\bar{a}_r^q(M)\right)^{\frac{1}{q}}\right)^q\right)^{\frac{1}{q}}+\|a\|_{n_{p,q}}$$

$$=\left(\sum_{r=1}^{\infty}r^{q-1}\bar{a}_r^q(M)\sum_{m=\log_2 r}^{\infty}2^{-\theta mq}\right)^{\frac{1}{q}}+\|a\|_{n_{p,q}(M)}$$

$$\asymp\left(\sum_{r=1}^{\infty}r^{(1-\theta)q-1}\bar{a}_r^q(M)\right)^{\frac{1}{q}}+\|a\|_{n_{p,q}(M)}=c\|a\|_{n_{p,q}(M)}.$$

Hence,

$$\|a\|_{(n_{1,\infty}(M),n_{\infty,\infty}(M))_{\theta,q}}\lesssim\|a\|_{n_{p,q}}.$$

So we get the embedding

$$n_{p,q}(M) \hookrightarrow (n_{1,\infty}(M), n_{\infty,\infty}(M))_{\theta,q},$$

where $\frac{1}{p} = 1 - \theta,\ \theta \in (0, 1)$.

Let us now prove the reverse embedding. Let $k \in \mathbb{N}$, $a \in (n_{1,\infty}, n_{\infty,\infty})_{\theta,q}$ and $a = a_0 + a_1$ be an arbitrary representation, where $a_0 \in n_{1,\infty}(M)$ and $a_1 \in n_{\infty,\infty}(M)$. Obviously, $\bar{a}_k(M) \leqslant \bar{a}_k^0(M) + \bar{a}_k^1(M)$. Then, if we denote $v(t) = t$, where $t \in (1, \infty)$, then

$$\sup_{v \geqslant k} k\bar{a}_k(M) \leqslant \sup_{k>0} k\bar{a}_k^0(M) + \sup_{v(t)\geqslant k} k\bar{a}_k^1(M) \leqslant \sup_{k\geqslant 1} k\bar{a}_k^0(M) + t \sup_{k\geqslant 1} \bar{a}_k^1(M).$$

Taking into account the arbitrariness of the representation $a = a_0 + a_1$, we have

$$\sup_{v\geqslant k>0} k\bar{a}_k(M) \leqslant K(t, a; n_{1,\infty}(M), n_{\infty,\infty}(M)).$$

Therefore, for $0 < q \leqslant \infty$, we have

$$\int_0^\infty \left(t^{-\theta} K(t, a; n_{1,\infty}(M), n_{\infty,\infty}(M))\right)^q \frac{dt}{t} \geqslant \int_1^\infty \left(t^{-\theta} \sup_{v\geqslant k>0} k\bar{a}_k(M)\right)^q \frac{dt}{t}$$

$$\geqslant c_1 \int_1^\infty \left(t^{-\theta} \sup_{t\geqslant k>0} k\bar{a}_k(M)\right)^q \frac{dt}{t} \geqslant c_2 \sum_{r=0}^\infty \left(2^{-\theta r} \sup_{2^r\geqslant k>0} k\bar{a}_k(M)\right)^q$$

$$\geqslant c_2 \sum_{r=1}^\infty \left(2^{r/p} \bar{a}_{2^r}(M)\right)^q.$$

Thus we get the embedding

$$(n_{1,\infty}(M), n_{\infty,\infty}(M))_{\theta,q} \hookrightarrow n_{p,q}(M), \tag{3}$$

where $\frac{1}{p} = 1 - \theta,\ \theta \in (0, 1)$.

Hence the relation (2) holds. To prove the general case, we use the reiteration theorem [4, Theorem 3.5.3].

Let $1 < p_0 < p_1 < \infty$. From (2) it follows that there are $\theta_0, \theta_1 \in (0, 1)$ such that

$$\begin{aligned} (n_{1,\infty}(M), n_{\infty,\infty}(M))_{\theta_0,q_0} &= n_{p_0,q_0}(M), \\ (n_{1,\infty}(M), n_{\infty,\infty}(M))_{\theta_1,q_1} &= n_{p_1,q_1}(M), \end{aligned} \tag{4}$$

then by the reiteration theorem, it follows that

$$(n_{p_0,q_0}(M), n_{p_1,q_1}(M))_{\theta,q} = (n_{1,\infty}(M), n_{\infty,\infty}(M))_{\eta,q} = n_{p,q}(M).$$

In the last equality, we took into account that $\eta = (1-\theta)\theta_0 + \theta\theta_1$.
This proves the theorem. □

3 As a corollary, an interpolation theorem of Marcinkevich type is presented.

Corollary 1 *Let* $2 \leqslant p_0 < p_1 < \infty$, $1 \leqslant q_0, q_1 < \infty$, $q_0 \neq q_1$, $0 < \tau, \sigma < \infty$, M *is the set of all segments from* $\mathbb{Z}$, $G = \{a = \{a_k\}_{k\in\mathbb{Z}},\ a_k \geqslant 0\}$. *If the following inequalities hold for a quasilinear operator:*

$$\|Ta\|_{n_{q_0,\infty}(M)} \leqslant F_0\|a\|_{n_{p_0,\sigma}(M)},\quad a \in n_{p_0,\sigma}(M), \tag{5}$$

$$\|Ta\|_{n_{q_1,\infty}(M)} \leqslant F_1\|a\|_{n_{p_1,\sigma}(M)},\quad a \in n_{p_1,\sigma}(M), \tag{6}$$

then for any $a \in G \cap n_{p,\tau}$ *we have*

$$\|Ta\|_{n_{q,\tau}(M)} \leqslant cF_0^{1-\theta}F_1^{\theta}\|a\|_{n_{p,\tau}(M)}, \tag{7}$$

where $\frac{1}{p} = \frac{1-\theta}{p_0} + \frac{\theta}{p_1}$, $\theta \in (0,1)$ *and the corresponding constant* c *depends only on* $p_i, q_i, \sigma, i = 0, 1$.

Proof According to the real interpolation method [4, Theorem 3.1.2] and the inequalities (5) and (6), it follows

$$\|Ta\|_{(n_{q_0,\infty}(M),n_{q_1,\infty}(M))_{\theta,\tau}} \leqslant F_0^{1-\theta}F_1^{\theta}\|a\|_{(n_{p_0,\sigma}(M),n_{p_1,\sigma}(M))_{\theta,\tau}}.$$

From the relation (3) we have that

$$\|Ta\|_{n_{q,\tau}(M)} \leqslant c\|Ta\|_{(n_{q_0,\infty}(M),n_{q_1,\infty}(M))_{\theta,\tau}}.$$

From Theorem 1, taking into account that $a = \{a_k\}_{k\in\mathbb{Z}},\ a_k \geqslant 0$, we get

$$\|a\|_{n_{p,\tau}(M)} \asymp \|a\|_{(n_{p_0,\sigma}(M),n_{p_1,\sigma}(M))_{\theta,q}}.$$

□

Acknowledgments This research was supported by the Ministry of Education and Science of the Republic of Kazakhstan (project no.AP14870758).

References

1. R. Akylzhanov, M. Ruzhansky, $L_p - L_q$ multipliers on locally compact groups. J. Fun. Anal. **278**(3), 108324, 1–49 (2020)
2. R. Akylzhanov, M. Ruzhansky, Net spaces on lattices, Hardy-Littlewood type inequalities, and their converses. Eurasian Math. J. **8**(3), 10–27 (2017)
3. R. Akylzhanov, M. Ruzhansky, E.D. Nursultanov, Hardy-Littlewood, Hausdorff-Young-Paley inequalities, and $L_p - L_q$ Fourier multipliers on compact homogeneous manifolds. J. Math. Anal. Appl. **479**(2), 1519–1548 (2019)
4. J. Bergh, J. Löfström, *Interpolation Spaces. An Introduction* (Springer, Berlin, 1976)
5. K.A. Bekmaganbetov, E.D. Nursultanov, On interpolation and embedding theorems for the spaces $B_{pr}^{\alpha q}([0, 2\pi)^n)$. Anal. Math. 35(3), 169–188 (2009)
6. K.A. Bekmaganbetov, E.D. Nursultanov, *On interpolation and embedding theorems for the spaces* $B_{p\tau}^{\alpha q}(\Omega)$. Math. Notes **84**(5), 733–736 (2008)
7. K.A. Bekmaganbetov, E.D. Nursultanov, Embedding theorems for anisotropic Besov spaces $B_{pr}^{\alpha q}([0, 2\pi)^n)$. Izv. Math. **73**(4), 655–668 (2009)
8. K.A. Bekmaganbetov, E.D. Nursultanov, Interpolation of spaces $B_{pr}^{\alpha q}(R^n)$ and $F_{pr}^{\alpha q}(\mathbb{R}^n)$ and applications. Proc. Inst. Math. Natl. Acad. Sci. Ukraine **5**(3), 34–46 (2008)
9. O. Blasco, A. Ruiz, L. Vega, Non interpolation in Morrey-Campanato and block spaces. Ann. Scuola Norm. Sup. Pisa Cl. Sci. **4**, 31–40 (1999)
10. E.D. Nursultanov, Net spaces and inequalities of Hardy-Littlewood type. Sb. Math. **189**(3), 399–419 (1998)
11. E.D. Nursultanov, On the coefficients of multiple Fourier series in L_p - spaces. Izv. Math. **64**(1), 93–120 (2000)
12. E.D. Nursultanov, T.U. Aubakirov, Interpolation methods for stochastic processes spaces. Abstr. Appl. Anal. **2013**, 1–12 (2013)
13. E.D. Nursultanov, A.G. Kostyuchenko, Theory of control of "catastrophes". Russ. Math. Surv. **53**(3), 628–629 (1998)
14. E.D. Nursultanov, N.T. Tleukhanova, Lower and upper bounds for the norm of multipliers of multiple trigonometric Fourier series in Lebesgue spaces. Func. Anal. Appl. **34**(2), 151–153 (2000)
15. E.D. Nursultanov, S. Tikhonov, Net spaces and boundedness of integral operators. J. Geom. Anal. **21**, 950–981 (2011)
16. A. Ruiz, L. Vega, Corrigenda to "Unique continuation for Schrodinger operators" and a remark on interpolation of Morrey spaces. Publicacions Matem'atiques **39**, 405–411 (1995)

Titchmarsh Theorem by Modulus of Continuity for the Deformed Hankel–Lipschitz Classes

Vishvesh Kumar and Joel E. Restrepo

Abstract We give an analogue of the Titchmarsh theorem by modulus of continuity for the deformed Hankel-Lipschitz classes involving a deformed Hankel translation operator and equivalently described by the growth of the norm of the deformed Hankel transform for functions in $L^p(\mathbb{R}, |x|^{2\alpha-1}dx)$ ($\alpha > 1/4$, $1 \leqslant p \leqslant 2$).

2010 Mathematics Subject Classification 26A16, 42A38, 46E15

1 Introduction

Titchmarsh in [32, Theorem 85] gave a characterisation of Hölder Lipschitz classes in terms of the asymptotic estimate of the growth of the norm of the Fourier transform. This idea was extended by Younis in [33] using the growth of functions $(t^\alpha(\log t)^\gamma$ $(0 < \alpha \leqslant 1, \gamma \in \mathbb{R})$ as $t \to +0)$ to define the Lipschitz class. In the recent paper [11], it was extended to a more general family of growths called modulus (moduli) of continuity. Therefore, similar results can be expected to be proved in other settings. Different researchers have then written Lipschitz classes by some suitable generalised translation operators, which somehow balance the growth of several transforms with similar behaviour to the Fourier one. These types of works can be, e.g. found in [7–9, 18, 31] or in more general settings in [10, 17, 25, 27].

Several generalisations of the Fourier transform have been studied by many researchers. For instance, we can mention the Dunkl transform which generalises the classical Fourier transform. It was first introduced by C. F. Dunkl in his seminal paper [16] around 30 years ago. Dunkl analysis was first introduced in [16] and further developed in, e.g. [1, 2, 14, 15, 20, 24, 28] and the references therein.

V. Kumar · J. E. Restrepo (✉)
Department of Mathematics: Analysis, Logic and Discrete Mathematics, Ghent University, Ghent, Belgium
e-mail: Vishvesh.Kumar@ugent.be; Joel.Restrepo@ugent.be

M. Ruzhansky, B. Torebek (eds.), *Extended Abstracts MWCAPDE 2023*, Research Perspectives Ghent Analysis and PDE Center 1, https://doi.org/10.1007/978-3-031-41665-1_9

Recently, in [4], it was introduced a generalisation of the Fourier transform and Dunkl transform, the so-called (k, a)-generalised Fourier transform. The variation of the tuple (k, a) gives the possibility to obtain different transforms, for example, the euclidean Fourier transform for $k = 0, a = 2$, the Dunkl transform for $k > 0, a = 2$ and a new unitary operator $\mathcal{F}_{k,1}$ for $k > 0, a = 1$. The $\mathcal{F}_{k,1}$ transform is called the deformed Hankel transform (or k-Hankel transform or $(k, 1)$-generalised Fourier transform) and has a rich structure, and many aspect of this transform has been extensively studied in [5, 6, 12, 19, 21, 24, 26, 30].

In this paper, we consider the deformed Hankel–Lipschitz classes defined by the deformed Hankel translation operator [5] and whose behaviour is controlled by a generalised modulus of continuity. We show that functions in the latter classes can be equivalently described by an asymptotic estimate for the growth of the norm of their deformed Hankel transform.

2 Preliminaries

Let us begin by recalling briefly some known definitions and facts about modulus of continuity and Dunkl analysis.

2.1 *Modulus of Continuity*

Definition 2.1 A non-negative function $\omega : I \to [0, +\infty)$ defined on a real interval I is called almost increasing (almost decreasing) if there exists a constant $C \geqslant 1$ such that $\omega(t) \leqslant C\omega(s)$ for all $t, s \in I$ with $t \leqslant s$ $(t \geqslant s)$.

Definition 2.2 For a fixed $\delta_0 \in (0, +\infty)$, the function $\omega : [0, \delta_0] \to [0, +\infty)$ is called a *modulus of continuity* if:

(1) $\omega(0) = 0$ and $\omega(t)$ is continuous on $[0, \delta_0]$.
(2) $\omega(t)$ is almost increasing on $t \in [0, \delta_0]$.
(3) $\frac{\omega(t)}{t}$ is almost decreasing on $t \in [0, \delta_0]$.

We endow the modulus of continuity considered here with the following so-called Zygmund type conditions:

- There exists a constant C such that for all $t \in [0, \delta_0]$,

$$\int_0^t \frac{\omega(x)}{x}\,\mathrm{d}x \leqslant C\omega(t). \tag{1}$$

- There exists a constant C such that for all $t \in [0, \delta_0]$,

$$\int_t^{\delta_0} \frac{\omega(x)}{x^2}\,\mathrm{d}x \leqslant C\frac{\omega(t)}{t}. \tag{2}$$

A modulus of continuity ω satisfying the Zygmund-type conditions (1) and (2) belongs to the Bari–Stechkin classes [3, 22, 23, 29]. Classical examples of this class are t^γ, $t^\gamma\left(\ln\frac{1}{t}\right)^\lambda$ and $t^\gamma\left(\ln\ln\frac{1}{t}\right)^\lambda$, where $\gamma \in (0, 1)$ and $\lambda \in \mathbb{R}$.

2.2 The Deformed Hankel Theory on $\mathbb{R}$

Let us collect some basic notation and results of the so-called deformed Hankel theory and α-Hankel transform on $\mathbb{R}$; see, e.g. [4, 6].

Let $L_{p,\alpha} := L^p(\mathbb{R}, |x|^{2\alpha-1})$ $(1 \leqslant p < +\infty, \alpha > 1/4)$ be the space of functions from L^p defined on $\mathbb{R}$ endowed with the norm

$$\|f\|_{p,\alpha} = \left(c_\alpha \int_{\mathbb{R}} |f(x)|^p |x|^{2\alpha-1} \mathrm{d}x\right)^{1/p}, \quad c_\alpha = \frac{1}{2\Gamma(2\alpha)}.$$

Similarly, we can also define $L_{\infty,\alpha}$:

$$\|f\|_{\infty,\alpha} = \underset{x\in\mathbb{R}}{\text{ess-sup}}|f(x)|.$$

The deformed Hankel transform $\mathcal{F}_\alpha(f)$, $\alpha > 1/4$ for $f \in L_{1,\alpha}$ is defined by

$$\mathcal{F}_\alpha(f)(\lambda) = c_\alpha \int_{\mathbb{R}} f(x) B_\alpha(\lambda x)|x|^{2\alpha-1} \mathrm{d}x, \quad \lambda \in \mathbb{R},$$

where the deformed Hankel kernel $B_\alpha(\lambda, x) = B_\alpha(\lambda x)$ is given by

$$B_\alpha(\lambda x) = j_{2\alpha-1}(2|\lambda x|^{1/2}) - \frac{\Gamma(2\alpha)}{\Gamma(2\alpha+2)} \lambda x j_{2\alpha+1}(2|\lambda x|^{1/2}),$$

and $j_\alpha(x)$ is the normalised Bessel function of the first kind of order α:

$$j_\alpha(x) = \Gamma(1+\alpha) \sum_{k=0}^{+\infty} \frac{(-1)^k}{k!\Gamma(k+\alpha+1)} \left(\frac{x}{2}\right)^{2k}, \quad x \in \mathbb{R}.$$

Using the fact that $|B_\alpha(\lambda x)| \leqslant 1$, one can easily see that

$$\|\mathcal{F}_\alpha f\|_{\infty,\alpha} \leqslant \|f\|_{1,\alpha}. \tag{3}$$

The kernel $B_\alpha(\lambda x)$ is associated with the solution [4, Theorem 5.7] of the following equation:

$$|x|\Delta_\alpha B_\alpha(\lambda x) + |\lambda| B_\alpha(\lambda x) = 0,$$

where Δ_α is the well-known Dunkl Laplacian [13] given by

$$\Delta_\alpha := \frac{\mathrm{d}^2}{\mathrm{d}x^2} + \frac{2\alpha}{x}\frac{\mathrm{d}}{\mathrm{d}x} - \frac{\alpha}{x^2}(1 - S),$$

where $Sf(x) := f(-x)$.

The deformed Hankel transform satisfies the following Plancherel identity:

$$\|\mathcal{F}_\alpha f\|_{2,\alpha} = \|f\|_{2,\alpha}. \tag{4}$$

The inverse deformed Hankel transform is defined as

$$f(x) := c_\alpha \int_{\mathbb{R}} \mathcal{F}_\alpha(f)(\lambda) B_\alpha(\lambda x)|\lambda|^{2\alpha-1}\mathrm{d}\lambda.$$

By the Plancherel identity (4) and the estimate (3) with the help of the Riesz–Thorin interpolation theorem, give the following Hausdorff–Young inequality for the deformed Hankel transform [6, Corollary 3.2]:

$$\|\mathcal{F}_\alpha(f)\|_{q,\alpha} \leqslant C\|f\|_{p,\alpha},$$

for any $f \in L_{\alpha,p}$ with $1 \leqslant p \leqslant 2$ and q such that $\frac{1}{p} + \frac{1}{q} = 1$, where the positive constant C does not depend on f. This inequality is one of the main ingredients to prove the main result. For more detail on the kernel $B_\alpha(\lambda x)$ and inequalities related to the deformed Hankel transform, see, e.g. [19, 21, 24].

The following translation operator will be used frequently in this paper [6, Formula (1.12)]:

$$T_h(f)(x) = \int_{\mathbb{R}} f(z) K_L(x, h, z)|z|^{2\alpha-1}\mathrm{d}z, \quad x, h \in [-\infty, +\infty],$$

and satisfies that

$$\mathcal{F}_\alpha(T_h f)(\lambda) = B_\alpha(\lambda h)\mathcal{F}_\alpha(f)(\lambda), \quad \text{for almost every } \lambda \in \mathbb{R},$$

and any $f \in L_{p,\alpha}$ $(1 \leqslant p \leqslant 2)$, $\alpha > 1/2$ and $h \in \mathbb{R}$ [6, Theorem 3.3]. The explicit form of the kernel K_L can be found in [6, Theorem 1.1]. It was proved in [6, Theorem 3.3] that T_h is a bounded operator on $L_{\alpha,p}$ for $1 \leqslant p \leqslant +\infty$.

3 Statements of the Results

We characterise some deformed Hankel–Lipschitz functions involving a deformed Hankel translation operator by an asymptotic estimate for the growth of the norm of the deformed Hankel transform, which in both cases use modulus of continuity.

We first introduce the generalised deformed Hankel–Lipschitz classes.

Definition 3.1 Let ω be a modulus of continuity. A function $f \in L_{p,\alpha}$ $(1 \leqslant p \leqslant 2, \alpha > 1/4)$ is said to be in the ω-deformed Hankel–Lipschitz class, denoted by $DLip(\omega, p)$, if there exists a constant $C > 0$ such that for all sufficiently small $h > 0$:

$$\|T_h f - f\|_{p,\alpha} \leqslant C\omega(h).$$

We endow the space $DLip(\omega, p)$ with the norm

$$\|f\|_{DLip(\omega,p)} = \|f\|_{p,\alpha} + \|f\|_{\#,DLip(\omega,p)},$$

$$\|f\|_{\#,DLip(\omega,p)} = \sup_{h\neq 0} \frac{\|T_h f - f\|_{p,\alpha}}{\omega(h)},$$

so that it becomes a Banach space.

Now, we are ready to present the main result of this paper. Everywhere below we assume that ω is a modulus of continuity.

Theorem 3.1 *The following assertions are true.*

(1) *Suppose that ω satisfies the condition (1) and $f \in L_{p,\alpha}$ with $1 \leqslant p \leqslant 2$ and $\alpha > 1/4$. If $f \in DLip(\omega, p)$, then there exists a constant $C > 0$ such that, for all sufficiently small $h > 0$, we have*

$$\int_{|\lambda|\geqslant \frac{1}{h}} |\mathcal{F}_\alpha(f)(\lambda)|^q |\lambda|^{2\alpha-1} \mathrm{d}\lambda \leqslant C\omega^q(h), \quad \frac{1}{p} + \frac{1}{q} = 1.$$

(2) *Suppose that ω satisfies the condition (2) and $f \in L_{2,\alpha}$ with $\alpha > 1/4$. We also assume that ω is bounded below on $[\delta_0, +\infty)$ by a positive number and that $\omega^2(t)/t^3 \in L^1\big([\delta_0, +\infty)\big)$. If there exists a constant $C > 0$ such that for all sufficiently small $h > 0$,*

$$\int_{|\lambda|\geqslant \frac{1}{h}} |\mathcal{F}_\alpha(f)(\lambda)|^2 |\lambda|^{2\alpha-1} \mathrm{d}\lambda \leqslant C\omega^2(h),$$

then $f \in DLip(\omega, 2)$.

The following corollary follows immediately from Theorem 3.1.

Corollary 3.1 *Let ω be a modulus of continuity satisfying the conditions (1) and (2). Assume that $\omega(t)$ is bounded below on $[\delta_0, +\infty)$ by a positive number and that $\omega^2(t)/t^3 \in L^1\big([\delta_0, +\infty)\big)$. Let us take $f \in L_{2,\alpha}$ with $\alpha > 1/4$. Then the following statements are equivalent:*

(1) *There exists a constant $C > 0$ such that for all sufficiently small $h > 0$,*

$$\|T_h f - f\|_{2,\alpha} \leqslant C\omega(h).$$

(2) *There exists a constant* $C > 0$ *such that for all sufficiently small* $h > 0$,

$$\int_{|\lambda|\geqslant\frac{1}{h}} |\mathcal{F}_\alpha(f)(\lambda)|^2|\lambda|^{2\alpha-1}\mathrm{d}\lambda \leqslant C\omega^2(h).$$

Acknowledgments The authors were supported by the FWO Odysseus 1 grant G.0H94.18N: Analysis and Partial Differential Equations, the Methusalem programme of the Ghent University Special Research Fund (BOF) (Grant number 01M01021) and FWO Senior Research Grant G011522N.

References

1. B. Amri, J.-Ph. Anker, M. Sifi, Three results in Dunkl analysis. Colloq. Math. **118**(1), 299–312 (2010)
2. J.-Ph. Anker, An introduction to Dunkl theory and its analytic aspects, in *Analytic, Algebraic and Geometric Aspects of Differential Equations*. Trends Math. (Birkhäuser/Springer, Cham, 2017), pp. 3–58
3. N.K. Bari, S.B. Stechkin, Best approximations and differential properties of two conjugate functions. (Russian) Trudy Mosk. Mat. Obshch. **5**, 483–522 (1956)
4. S. Ben Said, T. Kobayashi, B. Orsted, Laguerre semigroup and Dunkl operators. Compos. Math. **148**(4), 1265–1336 (2012)
5. S. Ben Said, L. Deleaval, Translation operator and maximal function for the $(k, 1)$-generalized Fourier transform. J. Funct. Anal. **279**(8), 108706 (2020). 32 pp
6. S. Ben Said, A product formula and a convolution structure for a k-Hankel transform on $\mathbb{R}$. J. Math. Anal. Appl. **463**(2), 1132–1146 (2018)
7. E.S. Belkina, S.S. Platonov, Equivalence of K-functionals and moduli of smoothness constructed by generalized Dunkl translations. Izv. Vyssh. Uchebn. Zaved. Mat. **8**, 3–15 (2008); Translation in Russian Math. (Iz. VUZ) **52**(8), 1–11 (2008)
8. W.O. Bray, M.A. Pinsky, Growth properties of Fourier transforms via moduli of continuity. J. Funct. Anal. **255**, 2265–2285 (2008)
9. W.O. Bray, Growth and integrability of Fourier transforms on Euclidean space. J. Fourier Anal. Appl. **20**, 1234–1256 (2014)
10. R. Daher, J. Delgado, M. Ruzhansky, Titchmarsh theorems for Fourier transforms of Hölder-Lipschitz functions on compact homogeneous manifolds. Monatsh. Math. **189**(1), 23–49 (2019)
11. R. Daher, A. Fernandez, J.E. Restrepo, Characterising extended Lipschitz type conditions with moduli of continuity. Results Math. **76**, 125 (2021)
12. H. De Bie, L. Pan, D. Constales, Explicit formulas for the Dunkl dihedral kernel and the (k, a)-generalized Fourier kernel. J. Math. Anal. Appl. **460**(2), 900–926 (2018)
13. C.F. Dunkl, Differential–difference operators associated to reflection groups. Trans. Am. Math. Soc. **311**, 167–183 (1989)
14. M.F.E. de Jeu, The Dunkl transform. Invent. Math. **113**, 147–162 (1993)
15. M. de Jeu, Paley-Wiener theorems for the Dunkl transform. Trans. Am. Math. Soc. **358**(10), 4225–4250 (2006)
16. C.F. Dunkl. Differential-difference operators associated to reflection groups. Trans. Am. Math. Soc. **311**, 167–183 (1989)
17. A. Fernandez, J.E. Restrepo, D. Suragan, Lipschitz and Fourier type conditions with moduli of 157 continuity in rank 1 symmetric spaces. Monatsh. Math. **197**, 353–364 (2022)

18. D. Gorbachev, S. Tikhonov, Moduli of smoothness and growth properties of Fourier transforms: two-sided estimates. J. Approx. Theory **164**(9), 1283–1312 (2012)
19. D.V. Gorbachev, V.I. Ivanov, S.Y. Tikhonov, Pitt's inequalities and uncertainty principle for generalized Fourier transform. Int. Math. Res. Not. IMRN **2016**(23), 7179–7200 (2016)
20. D.V. Gorbachev, V.I. Ivanov, S.Y. Tikhonov, Positive L^p-bounded Dunkl-type generalized translation operator and its applications. Constr. Approx. **49**(3), 555–605 (2019)
21. D.V. Gorbachev, V.I. Ivanov, S.Y. Tikhonov, On the kernel of the (k, a)-generalized Fourier transform, arxiv preprint, (2022). arXiv:2210.15730v2
22. A.I. Guseinov, H.Sh. Mukhtarov, *Introduction to the Theory of Nonlinear Singular Integral Equations* (Russian) (Nauka, Moscow, 1980)
23. V. Kokilashvili, A. Meskhi, H. Rafeiro, S. Samko, Integral operators in non-standard function spaces, in *Variable Exponent Lebesgue and Amalgam Spaces*, vol. 1 (Birkhäuser/Springer, Basel, 2016)
24. V. Kumar, M. Ruzhansky, $L^p - L^q$ boundedness of (k, a)-fourier multipliers with applications to nonlinear equations. Int. Math. Res. Not. **2023**(2), 1073–1093 (2023)
25. M. Kumar, V. Kumar, M. Ruzhansky, Titchmarsh theorems on Damek-Ricci spaces via moduli of continuity of higher order. https://arxiv.org/abs/2107.13044
26. S. Negzaoui, S. Oukili, Modulus of continuity and modulus of smoothness related to the deformed Hankel transform. Results Math. **76**(164), 1–17 (2021)
27. S.S. Platonov, The Fourier transform of functions satisfying the Lipschitz condition on rank 1 symmetric spaces. Sib. Math. J. **46**(6), 1108–1118 (2005)
28. M. Rösler, Generalized Hermite polynomials and the heat equation for Dunkl operators. Comm. Math. Phys. **192**, 519–541 (1998)
29. N. Samko, Weighted Hardy and singular operators in Morrey spaces. J. Math. Anal. Appl. **350**(1), 56–72 (2009)
30. W. Teng, Imaginary powers of $(k, 1)$-generalized harmonic oscillator. Complex Anal. Oper. Theory **16**(89), 1–21 (2022)
31. S. Tikhonov, Best approximation and moduli of smoothness: computation and equivalence theorems. J. Approx. Theory **153**, 19–39 (2008)
32. E.C. Titchmarsh, *Introduction to the theory of Fourier Integrals* (Oxford University Press, Amen House, London, 1948)
33. M.S. Younis, Fourier transforms of Dini–Lipschitz functions. Int. J. Math. Sci. **9**(2), 301–312 (1998)

Titchmarsh Theorems for Hölder-Lipschitz Functions on Profinite Groups

Juan Pablo Velasquez-Rodriguez

Abstract In this short note, we provide an extension to metrizable profinite groups of the classical theorems of Titchmarsh on the Fourier transform of Hölder-Lipschitz functions. This generalizes the results of Younis on compact zero-dimensional abelian groups to the noncommutative case and proves a relation between the Hölder-Lipschitz-continuity of functions and their "Sobolev regularity" given in terms of the Vladimirov-Taibleson operator. Since the class of profinite groups is fairly big, the formulation of our results requires to impose a special condition on the representation theory of the group. We prove that in particular such condition is satisfied by compact nilpotent metrizable profinite groups, which covers the case of compact nilpotent -adic Lie groups. In addition, we study the modulus of continuity of L^2-functions on the group, the functional spaces related to it, and its relation to the L^2-based Hölder-Lipschitz spaces. Finally, we also derive a characterisation for Dini-Lipschitz classes on metrizable profinite groups in terms of the behavior of their Fourier coefficients, together with a multiplier theorem for Hölder-Lipschitz-spaces.

2010 Mathematics Subject Classification Primary; 43A25, 58J40; Secondary: 20E18, 42B15, 42B25

In this note our main goal is to extend to noncommutative metrizable profinite groups the following two Titchmarsh theorems for Hölder-Lipschitz functions:

The author is supported by the FWO Odysseus 1 grant G.0H94.18N: Analysis and Partial Differential Equations.

J. P. Velasquez-Rodriguez (✉)
Department of Mathematics: Analysis, Logic and Discrete Mathematics, Ghent University, Ghent, Belgium

Departamento de Matematicas, Universidad del Valle, Cali, Colombia
e-mail: juanpablo.velasquezrodriguez@ugent.be; velasquez.juan@correounivalle.edu.co

M. Ruzhansky, B. Torebek (eds.), *Extended Abstracts MWCAPDE 2023*, Research Perspectives Ghent Analysis and PDE Center 1,
https://doi.org/10.1007/978-3-031-41665-1_10

First Titchmarsh Theorem *Let G be an abelian metrisable profinite group. Let $0 < \alpha \leq 1$ and $1 < p \leq 2$. If $f \in Lip_G(\alpha; p)$, then we have $\widehat{f} \in L^{\beta}(\widehat{G})$ for $\frac{p}{p+\alpha p-1} < \beta \leq q$, $1/p + 1/q = 1$. Here, the Lipschitz space $Lip_G(\alpha; p)$ is defined as*

$$Lip_G(\alpha; p) := \{f \in L^p(G) : \|f(h + \cdot) - f(\cdot)\|_{L^p(G)} = \mathcal{O}(|h|^{\alpha}) \text{ as } h \to 0\}, \quad 0 < \alpha \leq 1, \quad 1 < p \leq \infty.$$

Second Titchmarsh Theorem *Let $0 < \alpha \leq 1$ and $f \in L^2(G)$. Then $f \in Lip_G(\alpha; 2)$ if and only if*

$$\sum_{|\xi| \in \widehat{G} \setminus G_k^{\perp}} |\widehat{f}(\xi)|^2 = \mathcal{O}(|G/G_k|^{-2\alpha}),$$

for any $k \in \mathbb{N}_0$. Here $H^{\perp}$ denotes the annihilator of $H \leq G$ and $\{G_n\}_{n \in \mathbb{N}_0}$ in a strictly decreasing sequence of compact open subgroups that form a basis of neighborhoods at the identity.

These theorems, initially proven by Titchmarsh in $\mathbb{R}^n$, were later extended to compact abelian groups by M. Younis in his PhD dissertation [16]. Since then, numerous authors have attempted to replicate these findings in a variety of contexts where Fourier analysis is available. For instance, just to give some recent examples, R. Daher and collaborators have conducted a systematic exploration of Titchmarsh's results for transforms such as the Dunkl transform, Jacobi-Dunkl transform, and generalized Fourier-Bessel transform, among others. See [3–6] for further details. In recent years, M. El Hamma and A. Mahfoud treated the L^p-case for the first Hankel-Clifford transform [7], while H. Monaim and S. Fahlaoui investigated Titchmarsh-type theorems for the quaternionic Fourier transform [9].

In the totally disconnected case, S. Platonov extended Titchmarsh theorems to locally compact totally disconnected groups, which are sometimes called zero-dimensional groups. He proved an analogue of the second Titchmarsh theorem, first on the field of -adic numbers $\mathbb{Q}$ and later on locally compact abelian Vilenkin groups [13, 14]. Here is a prime number greater than 2 that we fix for the rest of the paper. Our work, on the other hand, focuses on compact groups much like in [2]. When studying compact groups, one mainly considers two classes: locally connected compact groups, such as any compact real Lie group, and totally disconnected compact groups, such as metrizable profinite groups, like compact Lie groups over non-Archimedean local fields. Younis considered both types of groups in his doctoral thesis and proved how the Lipschitzness of a function is related to its Fourier coefficients when the group is abelian. However, in the noncommutative case things become more complicated, and the question of the relationship between the Lipschitzness of a function and its group Fourier transform is still interesting. For locally connected groups, Younis's results have already been improved upon in [2], where the authors extended Titchmarsh theorems to compact homogeneous manifolds, and in particular to compact Lie groups. That covers the locally connected case and gives an idea on what to expect on other classes of compact groups, but omits the totally disconnected case, which is the subject of this paper. We will focus on compact Vilenkin groups, a class of metrizable topological groups that includes most examples of profinite groups, including compact Lie groups over non-Archimedean local fields.

Definition 1 We say that a topological group G is a *compact Vilenkin group* if G is a profinite group endowed with a strictly decreasing sequence of compact open subgroups $\mathscr{G} := \{G_n\}_{n\in\mathbb{Z}}$ such that

(i) It holds

$$2 \leq \varkappa_n := |G_n/G_{n+1}| < \infty,$$

for every $n \in \mathbb{N}_0$.

(ii)

$$G = \bigcup_{n\in\mathbb{N}_0} G_n, \quad \text{and} \quad \bigcap_{n\in\mathbb{N}_0} G_n = \{e\}.$$

(iii) The sequence $\{G_n\}_{n\in\mathbb{N}_0}$ form a basis of neighborhoods at $e \in G$.

If we also have

$$\sup_{n\in\mathbb{N}_0} |G_n/G_{n+1}| < \infty,$$

we say that G is a bounded-order Vilenkin group. If $|G_n/G_{n+1}|$ is constant we say that G is a constant-order Vilenkin group.

Remark 2 Let μ_G the unique normalized Haar measure on G, and denote by $|A|$ the measure of $A \subset G$. The above definition is linked to the following definition of ultrametric: for $x, y \in G$ their associated distance is defined as

$$\varrho_{\mathscr{G}}(x, y) = |xy^{-1}|_{\mathscr{G}} := \begin{cases} 0 & \text{if } x = y, \\ |G_n| & \text{if } xy^{-1} \in G_n \setminus G_{n+1}. \end{cases}$$

In particular we write $|x|_{\mathscr{G}} := \varrho_{\mathscr{G}}(x, e)$. Note that this distance function relies on the selection of a particular sequence of compact open subgroups. However, for the purposes of this work, the choice of sequence is not significant. Thus we will use a fixed sequence of subgroups, resulting in a fixed ultrametric on the group.

Fourier analysis on Vilenkin groups is still nowadays an active area of research, especially from the point of view of Vilenkin systems. In recent years, G. Gát and U. Goginava studied the pointwise summability of Fourier-Vilenkin series [8]; G. Tepnadze, L. Persson, and coauthors explored in detail the properties of martingale

Hardy spaces and Vilenkin systems [10–12]; and M. Avdispahi, N. Memi, and F. Weisz investigated maximal functions, Hardy spaces, and Fourier multiplier theorems on unbounded Vilenkin groups [1].

The problem of the convergence and decay rate of Fourier coefficients is a well-studied topic in Fourier analysis. For locally connected groups, especially Lie groups, one can relate the smoothness of a function to the decay rate of its Fourier coefficients. For instance, for a smooth function on the unit circle $f \in C^k(\mathbb{T})$, we have $|\widehat{f}(j)| = o(|j|^{-k})$, and for $0 < \alpha \leq 1$ and f satisfying the Hölder condition of order α, we have $|\widehat{f}(j)| = \mathcal{O}(|j|^{-\alpha})$.

For Vilenkin groups, the notion of differentiability is given in terms of the Vladimirov-Taibleson operator. To some extent, the theory of this operator is parallel to that of the fractional Laplacian on compact Lie groups, especially because it is nicely diagonalized by the group Fourier transform, just as the Laplace-Beltrami operator. Among other things, we can use Sobolev spaces $H_2^k(G)$ associated with the Vladimirov-Taibleson operator D^k to study the decay of Fourier coefficients. We recall here the definition for the reader

$$D^k f(x) := C_k \int_G |y|_{\mathscr{G}}^{-k-1}(f(xy^{-1}) - f(x))d\mu_G(y), \;\; k > 0,$$

where C_k is a normalization constant that will be specified later, and the Sobolev space $H_2^k(G)$ is defined as

$$\{f \in L^2(\mu_G) : D^k f \in L^2(\mu_G)\}.$$

When $G = \mathbb{Z}$, or G is any compact abelian Vilenkin group, we have once again, just like for the toroidal case, that $f \in H_2^k(\mathbb{Z})$ implies $|\widehat{f}(j)| = o(|j|^{-k})$, and for $0 < \alpha \leq 1$ and f satisfying the Hölder condition of order α, we have $|\widehat{f}(j)| = \mathcal{O}(|j|^{-\alpha})$. See [16] for all the details.

Alternatively, we can investigate whether the Fourier transform belongs to some L^p-space. For example, the Hausdorff-Young inequality implies that $\widehat{f} \in L^q$ whenever $f \in L^p$ for $1 < p \leq 2$ and $1/p + 1/q = 1$, and this fact can be strengthened if we assume some additional properties on f. Titchmarsh explored this idea in the context of Lipschitz spaces on $\mathbb{R}^n$, and Younis adapted his ideas to the setting of compact abelian groups, producing the first and second Titchmarsh theorems for Fourier series.

In this work, we continue in the same direction by considering functions $f \in Lip_{\mathscr{G}}(\alpha; p)$ in the Lipschitz space over a compact noncommutative Vilenkin group.

Definition 3 Let G be a compact Vilenkin group. Let $0 < \alpha \leq 1$ and $1 \leq p \leq \infty$ be real numbers. We define the Lipschitz space $Lip_{\mathscr{G}}(\alpha; p)$ by

$$Lip_{\mathscr{G}}(\alpha; p) := \{f \in L^p(\mu_G) \; : \; ||f(h\cdot) - f(\cdot)||_{L^p(\mu_G)} = O(|h|_{\mathscr{G}}^{\alpha}) \text{ as } |h|_{\mathscr{G}} \to 0\}.$$

For a compact group G, let $Rep(G)$ denote the collection of all continuous, finite dimensional, unitary irreducible representations of G. Let $\widehat{G}$ the collection of all irreducible representations in $Rep(G)$. As it is to be expected, the extension of the first and second Titchmarsh theorems to the noncommutative case require some non-trivial adjustments. This is best illustrated in [2] where, in order to obtain Titchmarsh theorems for compact homogeneous manifolds G/K, it is necessary to handle the behavior of the Fourier coefficients of the difference function $f(h\cdot) - f(\cdot)$ when $|h| \to 0$. These coefficients are $\widehat{f}(\xi)(\xi(h) - I_{d_\xi})$, $[\xi] \in \widehat{G}$, so it all comes down to the behavior of the map $\xi(h) - I_{d_\xi}$ when $|h| \to 0$.

In the profinite case, the same issue needs to be addressed, but the situation is much more diverse. The unitary irreducible representations of a compact Vilenkin group G have a non-trivial kernel, which is a compact open subgroup, so $\xi(h) - I_{d_\xi}$ becomes identically zero when $|h|_{\mathscr{G}} \to 0$. And even when $\xi(h) - I_{d_\xi} \neq 0_{d_\xi}$, a very common situation is that 1 is eigenvalue of $\xi(h)$, which is troublesome if one wants to estimate $\| \widehat{f}(\xi)(\xi(h) - I_{d_\xi}) \|_{HS}$ from below. Thus, in order to avoid this obstacle, we introduce the following condition on the representation theory of the group.

Condition (A) *We say that a compact Vilenkin group G satisfies Condition (A) if there is a natural number with the following property: for all $k \in \mathbb{N}_0$ there are points $h_1, \ldots, h$ satisfying $|h_1|_{\mathscr{G}} = \ldots = |h|_{\mathscr{G}} = |G/G_k|^{-1}$ and*

$$|G/G_k|^q \langle \xi \rangle_{\mathscr{G}}^{-q} \| \widehat{f}(\xi) \|_{HS}^q \lesssim \sum_{i=1} \| \widehat{f}(\xi)(\xi(h_i) - I_{d_\xi}) \|_{HS}^q, \quad 1 \leq q < \infty,$$

for any unitary irreducible representation $[\xi]$ non-trivial on $G_k \setminus G_{k+1}$. Here the function $\langle \cdot \rangle_{\mathscr{G}} : Rep(G) \to \mathbb{R}$ is defined in the following way:

$$\langle \pi \rangle_{\mathscr{G}} := \begin{cases} 1 & \text{if } \pi \text{ is the identity representation;} \\ |G/G_k| & \text{if } [\pi] \in Rep_k(G), \; n \in \mathbb{N}, \end{cases}$$

where

$$G_k^{\perp} := \{[\pi] \in Rep(G) \; : \; \pi|_{G_k} = I_{d_\pi}\}, \quad Rep_k(G) := G_k^{\perp} \setminus G_{k-1}^{\perp}.$$

Remark 4 To understand the above condition, it helps a lot to think on the abelian case. If $G = \mathbb{Z}$ then the unitary irreducible representations have the form $\chi_\xi(h) := e^{2\pi i \{h\xi\}}$, where $\xi \in \mathbb{Q}/\mathbb{Z}$ and $\{\cdot\}$ denotes the -adic fractional part function. Let $|h| = {}^{-n}$ and $|\xi| = {}^{m}$. When $m \leq n$, $\chi_\xi(h) = 1$ and $\chi_\xi(h) - 1 = 0$. When $m > n$, we can choose an h such that $\chi_\xi(h)$ is a ${}^{m-n}$ root of unity, so

$$|\chi_\xi(h) - 1| \gtrsim \sin \pi \frac{a_h}{{}^{m-n}} \gtrsim {}^{-(m-n)} = |\xi|^{-1}|h|^{-1}, \quad 1 \leq a_h < {}^{m-n},$$

which implies $|\xi|^{-1}|h|^{-1}|\widehat{f}(\xi)| \lesssim |\chi_\xi(h) - 1||\widehat{f}(\xi)|$. In the non-commutative case, this kind of lower bound is not trivial, thus the formulation of Condition (A). Given a noncommutative representation, it is possible in general for $\xi(h)$ to have 1 as eigenvalue, so the estimate $|h|_{\mathcal{G}}^{-1}\langle\xi\rangle_{\mathcal{G}}^{-1}\|\widehat{f}(\xi)\|_{HS}^{q} \lesssim \|\widehat{f}(\xi)(\xi(h)-I_{d_\xi})\|_{HS}^{q}$ might not be true for a single h.

Condition (A) restricts a lot the class of compact Vilenkin groups for which our results apply, but it still includes many important examples. We prove in this work how any compact nilpotent Vilenkin group satisfies this condition, and in particular any compact nilpotent -adic Lie group does as well, where $= (\mathfrak{g})$ equals the index of nilpotency of the $\mathbb{Z}$-Lie algebra.

Now with Condition (A) defined, we are ready to state the main results of this paper. We proved for functions $f \in Lip_{\mathcal{G}}(\alpha; p)$ the following two noncommutative versions of the Titchmarsh theorems.

Theorem 5 (First Titchmarsh Theorem) *Let G be a compact Vilenkin group satisfying Condition (A). Let p, q be real number such that $1 < p \leq 2$ and $1/p + 1/q = 1$. Let $0 < \alpha \leq 1$ and let $\alpha < \gamma < \alpha + \frac{1}{q}$. If $f \in Lip_{\mathcal{G}}(\alpha; p)$ then*

$$\Big(\sum_{|G/G_k| < \langle\xi\rangle_{\mathcal{G}}} d_\xi^{q(\frac{2}{q}-\frac{1}{2})} ||\widehat{f}(\xi)||_{HS}^{q} \Big)^{1/q} = \mathcal{O}(|G/G_k|^{-\alpha}).$$

Consequently $\widehat{D^\gamma f} \in L^\beta(\widehat{G})$ for

$$\frac{q}{(\alpha-\gamma)q+1} = \frac{p}{(\alpha-\gamma)p+p-1} \leq \beta < q.$$

Moreover $\widehat{f} \in L^\beta(\widehat{G})$ for

$$\frac{q}{\alpha q+1} = \frac{p}{\alpha p+p-1} < \beta \leq q.$$

Theorem 6 (Second Titchmarsh theorem) *Let $0 < \alpha \leq 1$ and $f \in L^2(\mu_G)$. Then $f \in Lip_{\mathcal{G}}(\alpha; 2)$ if and only if*

$$\sum_{\langle\xi\rangle_{\mathcal{G}} > |G/G_k|} d_\xi ||\widehat{f}(\xi)||_{HS}^{2} = \mathcal{O}(|G/G_k|^{-2\alpha}) \ \text{ as } k \to \infty.$$

Remark 7 Notice that in the hypothesis of the above theorems, we did not assume that G is a bounded-order Vilenkin group.

Something worth remarking about the above results is that, in contrast with [2], it might seem like the dimension of the group does not appear in the above theorems. In fact, we would like to emphasize how this information is actually there, but it is hidden in the choice of ultrametric for the group.

To do so, let > 2 be a prime number and let G be a d-dimensional linear -adic Lie sub-group of $GL_m(\mathbb{Z})$, where $\mathbb{Z}$ denotes the compact ring of -adic integers. We assume for the sake of our arguments that G is nilpotent, so that it can be thought as the exponential image of a certain nilpotent $\mathbb{Z}$-Lie algebra $\mathfrak{g}$. By choosing a $\mathbb{Z}$-basis $\{X_1, \ldots, X_d\}$ for $\mathfrak{g}$, we can write

$$\mathfrak{g} = \mathbb{Z}X_1 + \ldots + \mathbb{Z}X_d.$$

In this way we can identify points $\mathbb{Z}^d$ with points in the group via exponential map:

$$(x_1, \ldots, x_d) := \mathbb{exp}(x_1 X_1 + \ldots + x_d X_d).$$

By endowing the group with the sequence of compact open subgroups $\mathscr{G} = \{G_n\}_{n\in\mathbb{N}_0}$, $G_n := \mathbb{exp}(^n\mathfrak{g})$, we can see that G is a compact Vilenkin group, and the associated ultrametric function $|\cdot|_{\mathscr{G}}$ satisfies

$$|(x_1, \ldots, x_d)|_{\mathscr{G}} = \|(x_1, \ldots, x_d)\|^d, \quad \text{where } \|(x_1, \ldots, x_d)\| := \max_{1\leq j\leq d} |x_j|,$$

which allow us to rewrite the Vladimirov-Taibleson operator in a more convenient form

$$\mathscr{D}^\alpha f(x) := \frac{1 - ^\alpha}{1 - ^{-(\alpha+d)}} \int_G \frac{f(xy^{-1}) - f(x)}{\|y\|^{\alpha+d}} d\mu_G(y).$$

The definition of the Lipchitz space over G also changes slightly.

Definition 8 Let G be a compact nilpotent linear -adic Lie group. Let $0 < \alpha \leq 1$ and $1 \leq p \leq \infty$ be real numbers. We define the Lipschitz space $Lip_G(\alpha; p)$ by

$$Lip_G(\alpha; p) := \{f \in L^p(\mu_G) \ : \ ||f(h\cdot) - f(\cdot)||_{L^p(\mu_G)} = O(\|h\|^\alpha) \text{ as } \|h\| \to 0\}.$$

Notice the differences between the definitions of $Lip_G(\alpha; p)$ and $Lip_{\mathscr{G}}(\alpha; p)$. Given any metrizable profinite group G, the metric topology induces the existence of a sequence of compact open subgroups that coincide with the metric balls of the topology. And conversely, given a sequence of compact open subgroups, we can always define an invariant ultrametric for the group as in Remark 2. For any metrizable profinite group, there are many choices of metric and sequence of subgroups, so in principle the Lipschitz space on G depends on such choice, which is why we defined first the spaces $Lip_{\mathscr{G}}(\alpha; p)$. However, as we discussed previously, when G is a d-dimensional compact -adic Lie group, there is a natural choice of ultrametric, i.e., the -adic norm inherited from $\mathbb{Z}^d$. With this norm we can give a more "intrinsic" definition for the Lipchitz space independent of the choice of compact open subgroups, and in this case we use the notation $Lip_G(\alpha; p)$.

With the adapted version of the Lipchitz space over G, the Titchmarsh theorems take the following form. Notice how this time the dimension of the group appears in a similar way as in [2].

Theorem 9 *Let G be a compact nilpotent -adic Lie group of dimension d. Let p, q be real number such that $1 < p \leq 2$ and $1/p + 1/q = 1$. Let $0 < \alpha \leq 1$ and let $\alpha < \gamma < \alpha + \frac{d}{q}$. If $f \in Lip_G(\alpha; p)$ then*

$$\sum_{^{-k} < \langle \xi \rangle_G} d_\xi^{q(\frac{2}{q} - \frac{1}{2})} ||\widehat{f}(\xi)||_{HS}^q = \mathcal{O}(^{-k\alpha q}).$$

Consequently $\widehat{\mathscr{D}^\gamma f} \in L^\beta(\widehat{G})$ for

$$\frac{qd}{(\alpha - \gamma)q + d} = \frac{d}{(\alpha - \gamma) + d - \frac{d}{p}} \leq \beta \leq q.$$

Moreover $\widehat{f} \in L^\beta(\widehat{G})$ for

$$\frac{qd}{\alpha q + d} = \frac{pd}{\alpha p + pd - d} < \beta \leq q.$$

Here $\langle \xi \rangle_G$ denotes the eigenvalue of the operator

$$\mathbb{D} := \left(\frac{1 - ^{-d}}{1 - ^{-(1+d)}} I + \mathscr{D}^1\right),$$

corresponding to the matrix entries of the representation $[\xi] \in \widehat{G}$.

Theorem 10 *Let $0 < \alpha \leq 1$ and $f \in L^2(\mu_G)$. Then $f \in Lip_G(\alpha; 2)$ if and only if*

$$\sum_{\langle \xi \rangle_G > ^k} d_\xi ||\widehat{f}(\xi)||_{HS}^2 = \mathcal{O}(^{-k2\alpha}) \text{ as } k \to \infty.$$

Remark 11 In order to illustrate better the meaning of Condition (A), let us consider the Heisenberg group $\mathbb{H}_d$ over $\mathbb{Z}$, in turn defined as

$$\mathbb{H}_d(\mathbb{Z}) = \left\{ \begin{bmatrix} 1 & x^t & z \\ 0 & I_m & y \\ 0 & 0 & 1 \end{bmatrix} \in GL_{d+2}(\mathbb{Z}) \ : \ x, y \in \mathbb{Z}^d, \ z \in \mathbb{Z} \right\}.$$

Any unitary irreducible representation $[\pi]$ of $\mathbb{H}_d$ has two possibilities: either it is trivial on the center $\mathcal{Z}(\mathbb{H}_d)$, or it is not and it is the induced representation of a central character. If $[\pi]$ is trivial in the center, then it descents to a representation of the abelian group $\mathbb{H}_d/\mathcal{Z}(\mathbb{H}_d) \cong \mathbb{Z}^{2d}$, so that π is of the form

$$\pi_{\xi,\eta}(x, y, z) := e^{2\pi i \{x\xi + y\eta\}}, \quad \xi, \eta \in (\mathbb{Q}/\mathbb{Z})^d.$$

If $[\pi]$ is non-trivial in the center, then the restriction of the representation to the center gives rise to a central character $e^{2\pi i\{z\lambda\}}$, and the representation $[\pi]$ is the induced representation by said character. Hence, the noncommutative representations of $\mathbb{H}_d$ are indexed by $\lambda \in \mathbb{Q}/\mathbb{Z}$, and they satisfy

$$\pi_\lambda(0, 0, z) = e^{2\pi i\{z\lambda\}} I_{d_\lambda}.$$

Let us denote by $\widehat{\mathbb{H}}_d$ the collection of all unitary irreducible representations of $\mathbb{H}_d$. Then we can write

$$\widehat{\mathbb{H}}_d = \{[\pi_{(\xi,\eta,\lambda)}] = [\pi_{\xi,\eta}] \otimes [\pi_\lambda] \ : \ \xi, \eta \in (\mathbb{Q}/\mathbb{Z})^d, \ \lambda \in \mathbb{Q}/\mathbb{Z}\}.$$

Now we claim that $\mathbb{H}_d$ satisfies Condition (A) with $= 2$. To see it, just fix $n \in \mathbb{N}_0$. Then we can always take $h_1 := (x_0, y_0, 0)$ and $h_2 := (0, 0, z_0)$ where

$$\|x_0\| = \|y_0\| = |z_0| = {}^{-n},$$

so that for any representation $\pi_{(\xi,\eta,\lambda)}$ with $\|(\xi, \eta, \lambda\| > {}^{n}$ it holds

$$\pi_{(\xi,\eta,\lambda)}(x_0, y_0, 0) = e^{2\pi i\{x_0\xi + y_0\eta\}} \pi_\lambda(x_0, y_0, 0), \quad \text{and}$$
$$\pi_{(\xi,\eta,\lambda)}(0, 0, z_0) = e^{2\pi i\{z_0\lambda\}} I_{d_\lambda}.$$

If $|\lambda| = 1$, then

$$|\pi_{(\xi,\eta,\lambda)}(x_0, y_0, 0) - 1| \gtrsim {}^{n}\|(\xi, \eta)\|^{-1} = {}^{n}\|(\xi, \eta, \lambda)\|^{-1}.$$

If $|\lambda| > 1$,

$$\|\widehat{f}(\xi, \eta, \lambda)(\pi_{(\xi,\eta,\lambda)}(0, 0, z_0) - I_{d_\lambda})\|_{HS} \gtrsim^{n} |\lambda|^{-1} \|\widehat{f}(\xi, \eta, \lambda)\|_{HS}$$
$$\geq {}^{n}\|(\xi, \eta, \lambda)\|^{-1}\|\widehat{f}(\xi, \eta, \lambda)\|_{HS},$$

for any $\widehat{f}(\xi, \eta, \lambda) \in \mathbb{C}^{d_\lambda \times d_\lambda}$. This implies that $\mathbb{H}_d(\mathbb{Z})$ satisfies Condition (A) since we can easily check that $\langle \pi_{(\xi,\eta,\lambda)} \rangle_G \asymp \|(\xi, \eta, \lambda)\|$.

For the full account of our investigation on Titchmarsh theorems on profinite groups, and the proofs of the previously announced results, see [15]

References

1. M. Avdispahić, N. Memić, F. Weisz, Maximal functions, Hardy spaces and Fourier multiplier theorems on unbounded Vilenkin groups. J. Math. Anal. Appl. **390**(1), 68–73 (2012).
2. R. Daher, J. Delgado, M. Ruzhansky, Titchmarsh theorems for Fourier transforms of Hölder-Lipschitz functions on compact homogeneous manifolds. Monatsh. Math. **189**(1), 23–49 (2019)

3. R. Daher, M. El Hamma, An analog of Titchmarsh's theorem of Jacobi transform. Int. J. Math. Anal. **6**, 975–981 (2012)
4. R. Daher, M. El Hamma, A. Akhlidj, A. Dahani, On estimates for the generalized Dunkl transform and Titchmarsh's theorem in the space $L^p_{\alpha,Q}(\mathbb{R})(1 < p \leq 2)$. Facta Univ. Ser. Math. Inform. **37**, 31–40 (2022)
5. R. Daher, M.E. Hamma, Generalization of Titchmarsh theorem for the Fourier transform in the space $L^2(\mathbb{R})$. Ganit **33**, 77–82 (2014)
6. M. El Hamma, R. Daher, Generalization of Titchmarsh's theorem for the Bessel transform in the space $L_{p,\alpha}(\mathbb{R}_+)$. Ann. Math. Sil. **26**, 55–59 (2013)
7. M. El Hamma, A. Mahfoud, Generalization of Titchmarsh's theorem for the first Hankel-Clifford transform in the space $L^p_\mu((0, +\infty))$. Probl. Anal. Issues Anal. **29**(11), 56–65 (2022)
8. G. Gát, U. Goginava, Pointwise strong summability of Vilenkin–Fourier series. Math. Notes **108**(3–4), 499–510 (2020)
9. H. Monaim, S. Fahlaoui, Titchmarsh's theorems for the general double-sided orthogonal 2d-plans split QFT. J. Anal. **31**(1), 229–241 (2022)
10. N. Nadirashvili, L.-E. Persson, G. Tephnadze, F. Weisz, Vilenkin–Lebesgue points and almost everywhere convergence for some classical summability methods. Mediterr. J. Math. **19**, 5 (2022)
11. L.-E. Persson, F. Schipp, G. Tephnadze, F. Weisz, An analogy of the carleson–hunt theorem with respect to vilenkin systems. J. Fourier Anal. Appl. **28**, 3 (2022)
12. L.-E. Persson, G. Tephnadze, P. Wall, On the nörlund logarithmic means with respect to Vilenkin system in the martingale Hardy space H 1. Acta Math. Hungar. **154**(2), 289–301 (2017)
13. S.S. Platonov, An analogue of the Titchmarsh theorem for the Fourier transform on locally compact Vilenkin groups. p-Adic Numbers Ultrametric Anal. Appl. **9**(4), 306–313 (2017)
14. S.S. Platonov, An analogue of the Titchmarsh theorem for the Fourier transform on the group of p-adic numbers. p-Adic Numbers Ultrametric Anal. Appl. **9**(2), 158–164 (2017)
15. J.P. Velasquez-Rodriguez, Titchmarsh Theorems for Hölder-Lipschitz functions on profinite groups (2023) arXiv e-prints, arXiv:2303.18073
16. M. Younis, Fourier transform of Lipschitz functions on compact groups. Ph.D. thesis, McMaster University, 1974

An Efficient Neural Network for the Diagnosis of Parkinson's Disease Using Dynamic Handwriting Analysis

Xuechao Wang, Sven Nõmm, Junqing Huang, Marianna Chatzakou, and Michael Ruzhansky

Abstract Dynamic handwriting analysis of patients has become an important auxiliary method for the early diagnosis and treatment of Parkinson's disease. In this paper, a novel and efficient hybrid neural network model is proposed to evaluate the potential of handwriting sequence information in identifying Parkinson's disease, in which the hybrid model combines the respective advantages of long short-term memory and one-dimensional convolutional neural networks. Extensive experiments on the public dataset confirm that the proposed hybrid model can effectively extract discriminative sequence information for efficient diagnosis of Parkinson's disease, where the hybrid neural network has only 0.084 MB and above 90% diagnostic accuracy, making it highly reliable and robust for future clinical applications.

1 Introduction

Parkinson's disease (PD) is one of the most widespread and disabling neurodegenerative disorders. The disease is due to the progressive loss of dopaminergic neurons in the middle brain, leading to a decrease in functional, cognitive, and behavioral

---

X. Wang (✉) · J. Huang · M. Chatzakou
Department of Mathematics: Analysis, Logic and Discrete Mathematics, Ghent University, Ghent, Belgium
e-mail: xuechao.wang@ugent.be; junqing.huang@ugent.be; marianna.chatzakou@ugent.be

S. Nõmm
Department of Software Science, School of Information Technologies, Tallinn University of Technology, Tallinn, Estonia
e-mail: elli.valla@taltech.ee; sven.nomm@taltech.ee

M. Ruzhansky
Ghent Analysis and PDE Center, Department of Mathematics: Analysis, Logic and Discrete Mathematics, Ghent University, Ghent, Belgium

School of Mathematical Sciences, Queen Mary University of London, London, UK
e-mail: michael.ruzhansky@ugent.be; m.ruzhansky@qmul.ac.uk

M. Ruzhansky, B. Torebek (eds.), *Extended Abstracts MWCAPDE 2023*,
Research Perspectives Ghent Analysis and PDE Center 1,
https://doi.org/10.1007/978-3-031-41665-1_11

abilities [1]. Currently, there is no cure, but if PD is diagnosed at an early stage, then progression of the disease can be significantly slowed with appropriate treatment methods [2, 3].

Handwriting is an extremely common but complex human activity in a variety of leisure and professional settings, which requires fine dexterity skills and involves an intricate blend of cognitive, sensory, and perceptual-motor components [4]. Changes in it have been well documented to be promising biomarkers for the diagnosis of early PD [5, 6], and once diagnosed, it allows for later neuroprotective interventions. In fact, a growing body of knowledge provides evidence that it is possible to automatically distinguish between unhealthy and healthy individuals using simple and easy-to-perform handwriting tasks [7–10]. Therefore, the development of handwriting-based decision support tools is desirable as it could provide noninvasive and low-cost solutions to support standard clinical assessments carried out by human experts.

Within this research direction, dynamic (online) systems based on a digital tablet-based device can be used for the diagnosis of PD. When designing such a system, a critical step is to extract the most suitable features to describe the unique handwriting pattern. Usually, some dynamic features, such as traditional kinematic features [11, 12] or less common features based on entropy and signal-to-noise ratio [8], are derived from raw time-series data acquired by digital devices and then fed into classical statistical learning classifiers. It is worth noting that to obtain a complete statistical representation of these dynamic features, it is usually necessary to calculate the single-valued features of the feature vector (including mean, median, standard deviation, etc.) [11, 12]. However, although this "holistic" method can help classifiers to find efficient decision boundaries in the feature space, it may lose relevant details because an arbitrarily long feature vector is compressed into a single value. Another approach is to automatically learn features using a deep learning model. Several works based on the convolutional neural network (CNN) have addressed the problem of automatically extracting features from static two-dimensional (2D) images, which is constructed from dynamic handwriting signals [9, 13]. Although this approach represents a powerful alternative to artificially designed features, it also provides only a static holistic view of the handwriting patterns studied. In recent years, hybrid systems based on the recurrent neural network (RNN) and CNN, although underexplored in this field, have shown breakthrough performance in traditionally challenging tasks [14, 15]. Unlike many researchers who compress raw data into single-valued features or reconstruct static 2D images, without undermining the importance of above works, we hope to directly exploit the raw dynamic handwriting signals to reveal an individual unique fine motor control pattern (handwriting impairments) and gain new insights into dynamic handwriting processes.

To this end, an efficient artificial intelligence-based approach is proposed, which uses a novel hybrid neural network model to diagnose early PD using dynamic features as input to the model. The designed model for handwriting signal classification can achieve extremely high performance in a very lightweight structure. Furthermore, the forward difference algorithm applied in data processing extracts the derived features that reflect the symptoms of PD handwriting impairments from the geometric variables of the raw signal, further improving the diagnostic performance of the model, while requiring very little processing time, making the whole operation very efficient. Therefore, the presented deep learning method is extremely efficient, as it has high diagnostic performance, high robustness, and minimal computational demands, which also opens possibilities for subsequent clinical development.

The paper is organized as follows. Section 2 provides the reader with the necessary information about the data. Sections 3 and 4 presents the research approach and the network architecture. Section 5 presents the main results of the current studies. The discussion of the results achieved, the limitations of the proposed approaches, and possible future directions are discussed in Sect. 6.

2 Materials

The public *PaHaW* dataset collects handwriting data from 37 PD patients and 38 age and gender-matched HC subjects [7, 11]. During the acquisition of *PaHaW* dataset, each subject is asked to complete eight handwriting tasks according to the prepared pre-filled template at a comfortable speed. A recent study [16] showed that the use of spiral drawing can result in a better identification of PD compared to other writing tasks. Therefore, we only included spiral drawing samples from the *PaHaW* dataset. Figure 1 shows the shape of the spiral task. Handwriting signals are recorded using a digitizing tablet overlaid with a blank sheet of paper (signals recorded using an Intuos 4M pen of frequency 200 Hz). For each data point, the dynamic signal captures seven sets of independent dynamic variables, including y-coordinate, x-coordinate, timestamp, button status, altitude, azimuth, and pressure.

3 Data Processing

Data processing presented in this paper involves pre-processing, enhancement, and segmentation. First, the raw handwriting signal is first normalized: the min-max normalization technique is used to linearly map each individual variable of the raw handwriting signal in the range from 0 to 1. Note also that the timestamp (t) variable is discarded in this work, as it is expressed as a single-valued feature rather than a time-dependent vector feature [7].

After the pre-processing procedure, we employ a forward difference algorithm to extract the variation of independent dynamic variables in adjacent moments. In addition, we use matrix operations to further optimize the speed of the entire

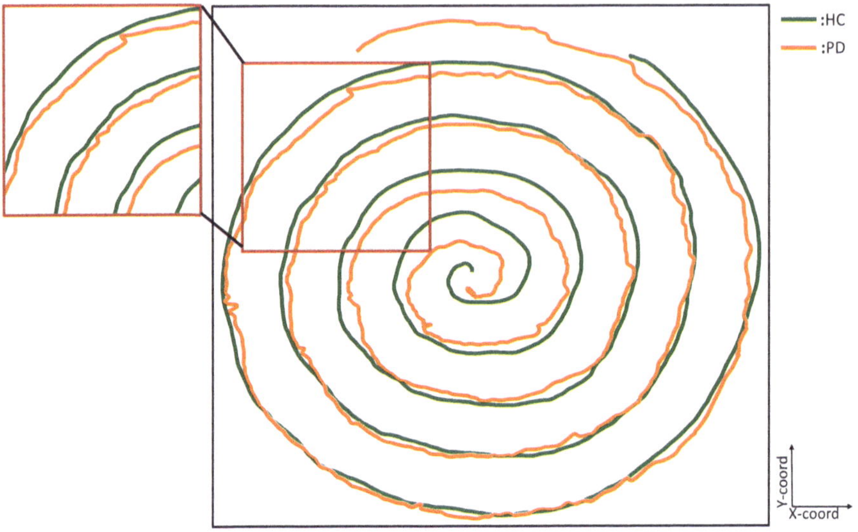

Fig. 1 Comparison of the hand drawings of Parkinson's disease (PD) patient and healthy control (HC) subject, where the relative positions of the hand drawings were reconstructed from the two-dimensional coordinates of successive hand drawing points as a function of time

operation, as in Eq. 1

$$\begin{bmatrix} \Delta x_1 \\ \Delta x_2 \\ \Delta x_3 \\ \vdots \\ \Delta x_m \end{bmatrix} = \begin{bmatrix} -1 & 1 & \cdots & 0 \\ 0 & -1 & \cdots & 0 \\ 0 & 0 & \cdots & 0 \\ \vdots & \vdots & \ddots & \vdots \\ 0 & 0 & \cdots & 1 \end{bmatrix} \begin{bmatrix} x_1 \\ x_2 \\ x_3 \\ \vdots \\ x_n \end{bmatrix} \tag{1}$$

where x_i is the data point at moment i of the raw variable; Δx_i is the data point at moment i of the extracted variable. Here, we only perform the forward difference operation on the x-coordinate and y-coordinate variables, while the azimuth, altitude, and pressure variables still adopt the raw values. Zero padding is used to ensure that all five individual dynamic variables are the same size.

Finally, a data segmentation technique is applied, using fixed-size window segments, known as temporal windows [17]. The enhanced signal is segmented into small patches of the same length with slight overlap to retain local temporal information, where the hyperparameters *window size* (w) and *stride size* (s) control the length of the temporal window and the degree of overlap, respectively. Therefore, the segmented patch data is a sequence $w \times 5$. Furthermore, the $w =$ 128 data point balances performance and efficiency under the proposed neural network architecture, and a window overlap of 50% is considered. Note that such a segmentation scheme generates more data samples, which is also crucial for subsequent model training.

4 LSTM-CNN

This section introduces our proposed hybrid neural network, whose overall network architecture is illustrated in Fig. 2. The proposed lightweight network architecture is a combination of long-short-term memory (LSTM) and one-dimensional (1D) CNN, which is called LSTM-CNN. The advantage of using 1D convolutions is twofold. First, 1D convolution reduces the overall cost of the computational model while satisfying the dimensional properties of the 1D handwriting signals. Second, 1D convolution can extract meaningful temporal dependencies from the input sequence to obtain higher-level representations. There are two convolutional layers; each layer proceeds with the 1D convolution operation, ReLU activation after convolution, and the activation function followed by the 1D max-pooling layer. Several filters of the same size are used in each layer to extract information across multiple time scales. The first applies 16 filters, with a kernel of size 3 and stride 2. The following layer involves a higher number of 32 filters with the same kernel of size 3 and stride 2. Finally, the model uses a fully connected layer to discriminate handwriting signals. The dropout layer temporarily removes nodes from the network with probability 0.5 during network training. On the other hand, the LSTM has a special set of memory units, where current input data and previous states affect the output of the next state, and the temporal features of the data can be extracted from the memory of historical information. Therefore, the combination of 1D convolutions

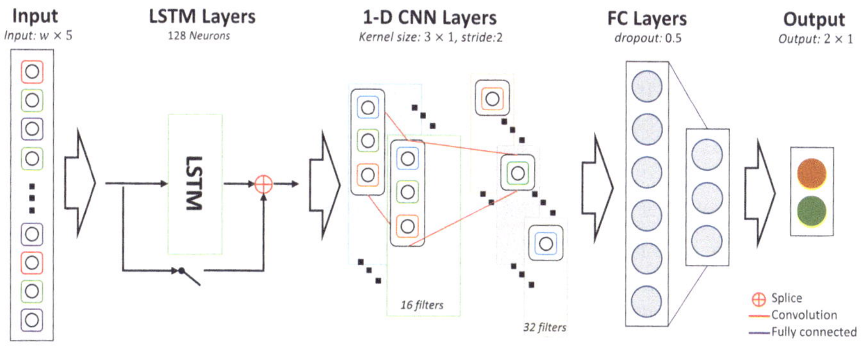

Fig. 2 Architectural details of the LSTM-CNN for the diagnosis of early PD. The input to the model is a patch sequence of length w, and feature dimension is a vector of 5 features. There are one LSTM layer, with 128 recurrent units. Next, there is a "shortcut connection" operation (called "Splice" layer for convenience), which concatenates the input data and the output of LSTM layers in the feature dimension. Two 1D convolutional layers with 16 and 32 filters and strides of 2 follow. The output layer has a single neuron with softmax to predict one of the two classes (PD or HC)

and LSTM is beneficial, especially when dealing with sequence-based dynamic data such as ours. First, an LSTM layer (128 hidden units) is stacked on top of the two previously mentioned 1D CNN layers; then a "Splice" layer (similar to the "shortcut connections" used in residual networks[18]) is added after the LSTM layer, which splices the original input with the output of the LSTM layer in *feature dimension* before feeding them into the 1D CNN layers to enrich the temporal features of the data; finally, to accommodate the input shape of the convolutional layer, the output of the "Splice" layer is expanded in dimension and can be represented as $(1, temporal\ windows, feature\ dimension)$. The overall network parameters are only 83.89 KB, and the floating-point operations are only 590.21 K.

5 Experiments

In this section, we report the results of a series of experiments that aimed to evaluate the effectiveness of the proposed method. The details of the experiments are described below.

Model training and testing were carried out on a PC that has an Intel(R) Core(TM) i7-11700K CPU with 3.60 GHZ(8 CPU), 32 GB RAM, and an NVIDIA GEFORCE RTX3070Ti graphics card with 8 GB memory. The initial learning rate was set at $1e-3$, and Adam [19] optimizer was used to train the model. The patches dataset was segmented into mini-batches of 64 data segments to optimize the training and validation processes. The cross-entropy loss function was used to optimize the model parameters. The hold-out method was used, and the average value of 5 experiments was used as the performance of the model.

To verify the generalization ability of our method, we conducted a series of experiments on *PaHaW*. Specifically, we applied the same data processing as described in Sect. 3, and the results obtained were averaged after repeating the experiment for 5 runs. In addition, we analyzed the previous literature with this particular database in order to contextualize our results. It is important to note that since different classifiers (SVM, RF, CNN, etc.) were used in these studies,

Table 1 Comparisons with state-of-the-art works on *PaHaW*

Author(s)	Feature set	Models	Outcomes (in %)	Year
Drotár et al. [11]	kinematic, spatio-temporal and pressure features	SVN,AB,KNN	Accuracy: 62.8	2016
Impedovo [8]	velocity-based features and classic features	SVM	Accuracy: 97.3	2019
Diaz et al. [9]	static images with dynamically enhanced	2D CNN+(SVM,RF)	Accuracy: 75.0	2019
Ours	dynamic handwriting sequence patterns	LSTM-CNN	Accuracy: 90.7	

reporting different classification accuracies, for a fair comparison, we only report here the best results in these studies. As can be inferred from Table 1, the sequence learning method based on 1D convolution and LSTM significantly outperforms the previously proposed method based on traditional dynamic features and 2D convolution on the same dataset. This confirms the effectiveness of the proposed method as a candidate solution for practical use in clinical settings. Furthermore, it is worth pointing out that the authors [8] used the non-nested validation scheme. However, it is well-known that selecting features a priori across the entire dataset can accidentally introduce bias into the classification process, which can lead to overly optimistic performance [20].

6 Conclusions

In this study, we propose a novel model based on LSTM and 1D CNN to identify unique fine motor control patterns in dynamic handwriting signals. Especially after the LSTM layer, a "shortcut connection" operation, similar to that in the residual network, is added, which significantly improves the model diagnosis rate while keeping the model efficient. Our experimental results confirm that the proposed technique is highly efficient compared to the state-of-the-art. A significant limitation of this study is the small size of the dataset we used, which may somewhat affect the generalizability of the results obtained. Despite these limitations, the reported performance values are indeed very promising, and the results of this study will hopefully make way for a working system in a clinical setting.

Acknowledgments This work in the project "ICT programme" was supported by the European Union through the European Social Fund. It was also partially supported by the FWO Odysseus 1 grant G.0H94.18N: Analysis and Partial Differential Equations and the Methusalem programme of the Ghent University Special Research Fund (BOF) (Grant number 01M01021). Michael Ruzhansky is also supported by EPSRC grant EP/R003025/2.

References

1. O. Hornykiewicz, Biochemical aspects of Parkinson's disease. Neurology **51**(2 Suppl 2), S2–S9 (1998)
2. D.L. Murman, Early treatment of Parkinson's disease: opportunities for managed care. Am. J. Manag. Care **18**(7), S183 (2012)
3. R.A. Hauser, Early pharmacologic treatment in Parkinson's disease. Am. J. Manag. Care **16**, S100–S107 (2010)
4. E. Carmeli, H. Patish, R. Coleman, The aging hand. J. Gerontol. A: Biol. Sci. Med. Sci. **58**(2), M146–M152 (2003)
5. M. Thomas, A. Lenka, P. Kumar Pal, Handwriting analysis in Parkinson's disease: current status and future directions. Mov. Disord. Clin. Pract. **4**(6), 806–818 (2017)

6. I. Aouraghe, G. Khaissidi, M. Mrabti, A literature review of online handwriting analysis to detect Parkinson's disease at an early stage. Multimed. Tools Appl. **81**, 1–26 (2022)
7. P. Drotár, J. Mekyska, I. Rektorová, L. Masarová, Z. Smékal, M. Faundez-Zanuy, Analysis of in-air movement in handwriting: a novel marker for Parkinson's disease. Comput. Methods Prog. Biomed. **117**(3), 405–411 (2014)
8. D. Impedovo, Velocity-based signal features for the assessment of Parkinsonian handwriting. IEEE Signal Process. Lett. **26**(4), 632–636 (2019)
9. M. Diaz, M.A. Ferrer, D. Impedovo, G. Pirlo, G. Vessio, Dynamically enhanced static handwriting representation for Parkinson's disease detection. Pattern Recogn. Lett. **128**, 204–210 (2019)
10. C.R. Pereira, S.A. Weber, C. Hook, G.H. Rosa, J.P. Papa, Deep learning-aided Parkinson's disease diagnosis from handwritten dynamics, in *2016 29th SIBGRAPI Conference on Graphics, Patterns and Images (SIBGRAPI)* (IEEE, 2016), pp. 340–346
11. P. Drotár, J. Mekyska, I. Rektorová, L. Masarová, Z. Smékal, M. Faundez-Zanuy, Evaluation of handwriting kinematics and pressure for differential diagnosis of Parkinson's disease. Artif. Intell. Med. **67**, 39–46 (2016)
12. E. Valla, S. Nomm, K. Medijainen, P. Taba, A. Toomela, Tremor-related feature engineering for machine learning based Parkinson's disease diagnostics. Biomed. Signal Process. Control **75**, 103551 (2022)
13. C.R. Pereira, D.R. Pereira, G.H. Rosa, V.H. Albuquerque, S.A. Weber, C. Hook, J.P. Papa, Handwritten dynamics assessment through convolutional neural networks: an application to Parkinson's disease identification. Artif. Intell. Med. **87**, 67–77 (2018)
14. Q. You, H. Jin, Z. Wang, C. Fang, J. Luo, Image captioning with semantic attention, in *Proceedings of the IEEE Conference on Computer Vision and Pattern Recognition* (2016), pp. 4651–4659
15. R. Chandra, U. Bhattacharya, A. Bera, D. Manocha, Traphic: trajectory prediction in dense and heterogeneous traffic using weighted interactions, in: *Proceedings of the IEEE/CVF Conference on Computer Vision and Pattern Recognition* (2019), pp. 8483–8492
16. A. Naseer, M. Rani, S. Naz, M.I. Razzak, M. Imran, G. Xu, Refining Parkinson's neurological disorder identification through deep transfer learning. Neural Comput. Applic. **32**, 839–854 (2020)
17. A. Jordao, A.C. Nazare Jr., J. Sena, W.R. Schwartz, Human activity recognition based on wearable sensor data: a standardization of the state-of-the-art. arXiv preprint arXiv:1806.05226
18. O. Russakovsky, J. Deng, H. Su, J. Krause, S. Satheesh, S. Ma, Z. Huang, A. Karpathy, A. Khosla, M. Bernstein et al., Imagenet large scale visual recognition challenge. Int. J. Comput. Vis. **115**, 211–252 (2015)
19. D.P. Kingma, J. Ba, Adam: a method for stochastic optimization. arXiv preprint arXiv:1412.6980
20. T. Hastie, R. Tibshirani, J.H. Friedman, J.H. Friedman, *The Elements of Statistical Learning: Data Mining, Inference, and Prediction*, vol. 2. (Springer, New York, 2009), pp. 1–758

Part II
Classical and Non-classical Problems of PDEs

On q^2-Analogue of the One-Dimensional Non-Homogeneous Heat Equation

Irfan Ali

Abstract In this paper, we discuss the q^2-analogue of non-homogeneous heat equation in one dimension. Namely, we consider non-homogeneous heat equation for Rubin's difference operator. Well-posedness results are presented in appropriate Sobolev-type spaces. In particular, we show that the heat equation generated by Rubin's difference operator have unique solutions. We even show that these solutions can be represented by explicit formulas. The presentation is based on the joint work with M. Ruzhansky and S. Shaimardan.

2010 Mathematics Subject Classification 33D05, 34C10, 39A10, 26D15

1 Introduction

One of the import equation in the theory of partial differential equations is heat equation. There are a lot of papers studying this equation with variety of modifications. Our main goal is to find the q^2-analogue of the following heat equation:

$$\begin{cases} u_t(t,x) - \Delta u(t,x) = f(t,x), & (t,x) \in \mathbb{R}^+ \times \mathbb{R}^n, \\ u(0,x) = \varphi(x), & x \in \mathbb{R}^n, \end{cases}$$

and Δ is the Laplace operator in $\mathbb{R}^n$.

We will use the q-difference operator which is defined by Rubin ([15]) instead of Δ. The history of q-calculus was started by the works of L. Euler [5] in 1748, while after almost a century, we see the work of E. Heine, in 1846. Later on in

I. Ali (✉)
Department of Mathematics: Analysis, Logic and Discrete Mathematics, Ghent University, Ghent, Belgium

Department of Mathematiccal Sciences, BUITEMS, Quetta, Pakistan

M. Ruzhansky, B. Torebek (eds.), *Extended Abstracts MWCAPDE 2023*,
Research Perspectives Ghent Analysis and PDE Center 1,
https://doi.org/10.1007/978-3-031-41665-1_12

the beginning of the twentieth century, the q-difference operator is introduced by F.H. Jackson [9] (see also [10]) and afterword a lot of more people like R.D. Carmichael [3], T.E. Mason [13], and C.R. Adams [1] (see also [2]) also studied the q-difference operator. In 1997, R.L. Rubin [14] using Rubin's operator (which is also a q-difference operator) investigated the spectral properties and introduced q-difference calculus associated with this Rubin's operator. He also studied the q^2-Fourier transform which is q^2-analogue of classical Fourier transform, while the exponential Fourier analysis was studied in [12]. q^2-analogue of Sobolov type spaces using the Rubin's q-difference operator has been studied by [16, 17]. By using the q^2-analogue of Fourier transform they also studied the solution of q-heat equation with Rubin's q-difference operator in [6, 18]. We study the Cauchy problem for the non-homogeneous heat equation generated by Rubin's difference operator $\partial_{q,x}$ in the Sobolev space: $W_q^2\left(\mathbb{R}_q^+\right)$

$$\begin{cases} D_{q^2,t}u(t,x) - \partial_{q,x}^2 u(t,x) = f(t,x), \ (t,x) \in \mathbb{R}^+ \times \mathbb{R}_q^+, \\ u(0,x) = \varphi(x), \qquad\qquad\qquad\qquad\qquad x \in \mathbb{R}_q^+. \end{cases}$$

We will give the uniqueness and show explicit formulas for the solution.

2 Preliminaries

We start by recalling some basic notation in the q-calculus; see, e.g., the books [11] and [4]. Throughout this paper, we assume for a fix q where $0 < q < 1$.

Let $\alpha \in \mathbb{R}$. Then a q-real number $[\alpha]_q$ is defined by

$$[\alpha]_q := \frac{1-q^\alpha}{1-q},$$

where $\lim\limits_{q\to 1} \frac{1-q^\alpha}{1-q} = \alpha$.

We introduce for any $x, a \in \mathbb{R}$

$$(x,a)_q^0 = 1, \quad (x,a)_q^n = \prod_{k=0}^{n-1}\left(x - q^k a\right), \quad (x,a)_q^\infty = \lim_{n\to\infty}(x,a)_q^n.$$

The q-analogues of the binomial coefficients are defined by

$$[n]_q! = \begin{cases} 1, & \text{if } n = 0 \\ [1]_q \times [2]_q \times \cdots \times [n]_q & \text{if } n \in \mathbb{N} \end{cases}$$

Similarly, the q-analogues of the gamma function is defined by

$$\Gamma_q(x) = \frac{(q,q)_q^\infty}{(q^x,q)_q^\infty}(1-q)^{1-x}.$$

Two q^2-analogues of the exponential function are given by (see [8])

$$e_{q^2}(z) = \sum_{k=0}^{\infty} \frac{1}{(q;q^2)_k} z^k = \frac{1}{(z;q^2)_\infty}, \ |z| < 1,$$

$$E_{q^2}(z) = \sum_{k=0}^{\infty} \frac{q^{k(k-1)}}{(q;q^2)_k} z^k = (-z;q^2)_\infty,$$

where E_{q^2} is entirely on complex domain.
In [7] they denote these q^2-analogue of exponential functions by

$$exp_{q^2}(z) = e_{q^2}((1-q)z) := \sum_{n=0}^{\infty} \frac{z^n}{[n]_{q^2}!},$$

$$Exp_{q^2}(z) = E_{q^2}((1-q)z) := \sum_{n=0}^{\infty} \frac{q^{n(n-1)}}{[n]_{q^2}!} z^n.$$

And also the q^2-exponential $e(z,q^2)$ (see [12, 14], and [15]) is defined by

$$e(z,q^2) = \cos(-iz;q^2) + i\sin(-iz;q^2)$$

where

$$\cos(z;q^2) = \sum_{k=0}^{\infty} \frac{(-1)^k q^{k(k+1)} z^{2k}}{[2k]_q!}$$

and

$$\sin(z;q^2) = \sum_{k=0}^{\infty} \frac{(-1)^k q^{k(k+1)} z^{2k+1}}{[2k+1]_q!}$$

are the q^2-trigonometric functions. The q^2-analogue differential operator is defined by (see [9])

$$D_{q^2} f(x) = \frac{f(x) - f(q^2 x)}{x(1-q^2)},$$

and the Rubin's q-difference operator is defined by (see[15])

$$\partial_q f(x) = \frac{f(q^{-1}x) + f(-q^{-1}x) - f(qx) + f(-qx) - 2f(-x)}{2x(1-q)}.$$

Moreover, we will use the q-partial differential operators $\partial_{q,x}u(t,x)$ in the following form:

$$\partial_{q,x}u(t,x) = \frac{u(t,q^{-1}x) + u(t,-q^{-1}x) - u(t,qx) + u(t,-qx) - 2u(t,-x)}{2x(1-q)}.$$

Note that if f is differentiable at x, then $\lim_{q\to 1} \partial_q f(x) = f'(x)$ and $\lim_{q\to 1} \partial_{q,x}u(t,x) = \frac{\partial u}{\partial x}(t,x)$.

The definite q-integral or the q-Jackson integral of a function f is defined by the formula (see [9] and [10])

$$\int_0^x f(t)d_qt := (1-q)x\sum_{k=0}^{\infty} q^k f(q^k x), \quad 0 < x < \infty, \tag{2.1}$$

and the improper q-integral of a function $f(x) : [0,\infty) \to \mathbb{R}$, is defined by the formula

$$\int_0^{\infty} f(t)d_qt := (1-q)\sum_{k=-\infty}^{\infty} q^k f(q^k),$$

Note that the series in the right-hand side of (2.1) converge absolutely.

The space $L_q^p\left(\mathbb{R}_q^+\right)$ for $\mathbb{R}_q^+ = \{q^k, k \in \mathbb{Z}\}$ is defined by ([16])

$$L_q^p\left(\mathbb{R}_q^+\right) = \left\{f : \left(\int_0^{\infty} |f(x)|^p d_qx\right)^{\frac{1}{p}} < \infty\right\},$$

for $0 < p < \infty$ and also the $L_q^{\infty}\left(\mathbb{R}_q^+\right)$ is defined by ([16])

$$L_q^{\infty}\left(\mathbb{R}_q^+\right) = \left\{f : \sup_{k\in\mathbb{Z}} |f(q^k)| < \infty\right\}.$$

Moreover, we use $S_q\left(\mathbb{R}_q^+\right)$ as the set of functions f defined on $\mathbb{R}_q^+$ satisfying the following condition:

$$P_{n,m,q}(f) = \sup_{x \in \mathbb{R}_q^+} \left| x^m \mathcal{D}_q^n f(x) \right| < \infty,$$

for all $m, n \in \mathbb{N}$ and where $\lim_{x \to 0} \mathcal{D}_q^n f(x)$ exists in $\mathbb{R}_q^+$ (see [7]).

Also the space $S'_q\left(\mathbb{R}_q^+\right)$ as the topological dual of $S_q\left(\mathbb{R}_q^+\right)$ defined by ([16]). The space $S'_q\left(\mathbb{R}_q^+\right)$ is called the space of tempered distributions on $\mathbb{R}_q^+$.

Definition 2.1 (See [7] and [15].) For $f \in S_q\left(\mathbb{R}_q^+\right)$ the q^2-Fourier transform $\hat{f}$ is defined as follows:

$$\mathcal{F}_q(f)(x) = \hat{f}(\xi) = K \int_0^\infty f(x) e(-ix\xi; q^2) d_q x, \quad \xi \in \mathbb{R}_q^+,$$

and also for any function $g \in S_q\left(\mathbb{R}_q^+\right)$, we have the following inverse transform $(\mathcal{F}^{-1}g)(x)$:

$$(\mathcal{F}^{-1}g)(x) = K \int_0^\infty e(ix\xi; q^2) g(\xi) d_q \xi, \quad x \in \mathbb{R}_q^+,$$

where

$$K = \frac{(1+q)^{\frac{1}{2}}}{\Gamma_{q^2}(\frac{1}{2})}. \tag{2.2}$$

for all $\xi \in \mathbb{R}_q^+$, where $n \in \mathbb{N}$.

Definition 2.2 (See [16].) For $s \in \mathbb{R}$, we use the Sobolev space $W_q^s\left(R_q^+\right)$ as

$$W_q^s\left(R_q^+\right) = \left\{ u \in S'_q\left(\mathbb{R}_q^+\right) : \left(1 + |\xi|^2\right)^{\frac{s}{2}} \widehat{u} \in L_q^2\left(\mathbb{R}_q^+\right) \right\},$$

equipped with the norms

$$\|u\|_{W_q^s(R_q^+)} = \left(\int_0^\infty \left(1 + |\xi|^2\right)^s |\widehat{u}(\xi)|^2 d_q \xi \right)^{\frac{1}{2}}.$$

Let $0 < T < \infty$. We use also the spaces $C^k\left([0,T]; W_q^s\left(\mathbb{R}_q^+\right)\right)$ and $C^k\left([0,T]; L_q^2\left(\mathbb{R}_q^+\right)\right)$ which their norms are defined ([16])respectively by

$$\|u\|_{C^k\left([0,T];W_q^s(\mathbb{R}_q^+)\right)} = \sum_{n=0}^{k} \max_{0\le t\le T} \|\partial_t^n u(t,.)\|_{W_q^s(R_q^+)}$$

and

$$\|u\|_{C^k\left([0,T];L_q^s(\mathbb{R}_q^+)\right)} = \sum_{n=0}^{k} \max_{0\le t\le T} \|\partial_t^n u(t,.)\|_{L_q^s(\mathbb{R}_q^+)}.$$

Lemma 2.1 *From [7, Proposition 4.4], we know the q^2-Fourier transformation of the solution $k(x,t;q)$ takes the following expression:*

$$\mathcal{F}_q[k(.,t;q)](x) = \hat{k}(.,t;q)(x) = K\exp_{q^2}(-tx^2), \tag{2.3}$$

where

$$k(x,t;q) = C(t,q)\exp_{q^2}\left(-qx^2/(1+q)^2 t\right), \quad t>0, \tag{2.4}$$

and

$$C(t,q) = \frac{(-q(1-q)/t(1+q), -qt(1+q)/1-q, q; q^2)_\infty}{2(1-q)\left(-q^2(1-q)/t(1+q), -t(1+q)/1-q, q^2; q^2\right)_\infty}.$$

Next, for $x, y \in \mathbb{R}_q^+$, we want to mention the Fourier multiplier operator (refer to [14, Section 3]) corresponding to translation by y (denoted by T_y) is

$$f(x\ominus y) = [T_y f](x) = K\int_{\mathbb{R}_q} e(-i\xi y; q^2)\hat{f}(\xi; q^2)e(i\xi x; q^2)d_q\xi, \tag{2.5}$$

whenever the integral makes sense.

3 Main Result

In the following theorem, we will show the solution of q^2-analogue of the non-homogeneous heat equation in one dimension. This result is obtained jointly with M. Ruzhansky and S. Shaimardan.

Problem 3.1 *We consider the following Cauchy problem:*

$$D_{q^2,t}u(t,x) - \partial^2_{q,x}u(t,x) = f(t,x), \quad x \in \mathbb{R}_q, t > 0 \tag{3.1}$$

$$u(0,x) = \varphi(x), \quad x \in \mathbb{R}_q, \tag{3.2}$$

where $f(t,x)$ and $\varphi(x)$ are known functions.

Theorem 3.2 *Assume that for a fix T and let $\varphi \in W^2_q\left(\mathbb{R}^+_q\right)$ and $f \in C\left([0,T]; W^2_q\left(\mathbb{R}^+_q\right)\right)$. Then there exists a unique solution of Problem (3.1)–(3.2):*

$$u \in C\left([0,T]; W^2_q\left(\mathbb{R}^+_q\right)\right) \cap C^1\left([0,T]; L^2_q\left(\mathbb{R}^+_q\right)\right).$$

The explicit formula of the solution can be written as

$$u(t,x) = \int_{\mathbb{R}_q} \varphi(y)k_1(x \ominus y, t, q)d_qy + \int_0^t K_\xi(t,s)d_qs,$$

for $(t,x) \in [0,T] \times \mathbb{R}^+_q$, where $k_1(x \ominus y, t, q) := Kk(x \ominus y, t; q)$, and $k(x,t,q)$ and K are given by (2.4) and (2.2), respectively, also

$$K_\xi(t,s) := K \int_{\mathbb{R}_q} \hat{f}(s,\xi)E_{q^2}(q^2\xi^2 s)e_{q^2}(-\xi^2 t)e(ix\xi; q^2)d_q\xi$$

where the estimates are

$$||u(t,.)||_{W^2_q(\mathbb{R}^+_q)} \lesssim \|\varphi\|_{W^2_q(\mathbb{R}^+_q)} + \|f\|_{C\left([0,T]; W^2_q(\mathbb{R}^+_q)\right)},$$

which also concludes that $\|u\|_{C\left([0,T]; W^2_q(\mathbb{R}^+_q)\right)} < \infty$.

Similarly

$$\|D_{q^2,t}u\|_{L^2_q(\mathbb{R}^+_q)} \lesssim \|\varphi\|_{W^2_q(R^+_q)} + \|f\|_{C\left([0,T]; W^2_q(\mathbb{R}^+_q)\right)}$$

thus we can conclude that $\|u\|_{C^1\left([0,T]; L^2_q(\mathbb{R}^+_q)\right)} < \infty$.

And our solution is unique.

Acknowledgments The research is financially supported by a grant the Methusalem programme of the Ghent University Special Research Fund (BOF) (Grant number 01M01021). Also author is also supported by Higher Education Commission of Pakistan, on Overseas Scholarship Program Phase-III.

References

1. C.R. Adams, On the linear ordinary q-difference equation. Ann. Math. (2) **30**(1–4), 195–205 (1928/29)
2. C.R. Adams, On the linear partial q-difference equation of general type. Trans. Am. Math. Soc. **31**(2), 360–371 (1929)
3. R.D. Carmichael, The general theory of linear q-difference equations. Am. J. Math. **34**(2), 147–168 (1912)
4. T. Ernst, *A Comprehensive Treatment of q-Calculus* (Birkhäuser/Springer Basel AG, Basel, 2012)
5. L. Euler, *Introductio in Analysin Infinitorum. Tomus Primus* (Sociedad Andaluza de Educación Matemática Thales, Seville; Real Sociedad Matemática Española, Madrid, 2000). Reprint of the 1748 original
6. A. Fitouhi, N. Bettaibi, K. Mezlini, On a q-analogue of the one-dimensional heat equation. Bull. Math. Anal. Appl. **4**(2), 145–173 (2012)
7. A. Fitouhi, R.H. Bettaieb, Wavelet transforms in the q^2-analogue Fourier analysis. Math. Sci. Res. J. **12**(9), 202–214 (2008)
8. G. Gasper, M. Rahman, Basic hypergeometric series, in *Encyclopedia of Mathematics and its Applications*, vol. 35 (Cambridge University Press, Cambridge, 1990). With a foreword by Richard Askey.
9. F.H. Jackson. XI.—on q-functions and a certain difference operator. Earth Environ. Sci. Trans. R. Soc. Edinb. **46**(2), 253–281 (1909)
10. F.H. Jackson, On q-definite integrals. Q. J. Pure Appl. Math. **41**(15), 193–203 (1910)
11. V. Kac, P. Cheung, *Quantum Calculus*. Universitext (Springer, New York, 2002)
12. T.H. Koornwinder, R.F. Swarttouw, On q-analogues of the Fourier and Hankel transforms. Trans. Am. Math. Soc. **333**(1), 445–461 (1992)
13. T.E. Mason, On properties of the solutions of linear q-difference equations with entire function coefficients. Am. J. Math. **37**(4), 439–444 (1915)
14. R.L. Rubin, A q^2-analogue operator for q^2-analogue Fourier analysis. J. Math. Anal. Appl. **212**(2), 571–582 (1997)
15. R.L. Rubin, Duhamel solutions of non-homogeneous q^2-analogue wave equations. Proc. Am. Math. Soc. **135**(3), 777–785 (2007)
16. A. Saoudi, A. Fitouhi, On q^2-analogue Sobolev type spaces. Matematiche (Catania) **70**(2), 63–77 (2015)
17. A. Saoudi, A. Fitouhi, Three applications in q^2-analogue Sobolev spaces. Appl. Math. E-Notes **17**, 1–9 (2017)
18. S. Shaimardan, L.-E. Persson, N. Tokmagambetov, Well-posedness of heat and wave equations generated by Rubin's q-difference operator in Sobolev spaces (2023). arXiv preprint arXiv:2301.07381

Non-local and Inverse Problems for the Rayleigh-Stokes Equation

Ravshan Ashurov and Oqila Mukhiddinova

Abstract This work consists of two parts. In the first part, a non-local boundary value problem is studied for the fractional version of the Rayleigh-Stokes equation known in hydrodynamics. Namely, instead of the initial condition, the condition $u(x, T) = \beta u(x, 0) + \varphi(x)$ is proposed, where β is an arbitrary real number. Note that if $\beta = 0$, then we get the backward problem, which is ill-posed. The main goal of this part of the work is to study the influence of parameter β on the correctness of the problem. The second part considers the homogeneous Rayleigh-Stokes equation and investigates the behavior of the solution of the Rayleigh-Stokes problem depending on the order of the fractional derivative α. In particular, it is proved that for sufficiently large t, the norm $||u(x, t)||_{L_2(\Omega)}$ of the solution decreases in α. Moreover, assuming the parameter α to be unknown, an additional condition on the solution of the problem is found that uniquely determines the order of the derivative α.

R. Ashurov (✉)
V.I. Romanovskiy Institute of Mathematics, Uzbekistan Academy of Science, Tashkent, Uzbekistan

School of Engineering, Central Asian University, Tashkent, Uzbekistan

O. Mukhiddinova
V.I. Romanovskiy Institute of Mathematics, Uzbekistan Academy of Science, Tashkent, Uzbekistan

Tashkent University of Information Technologies, Tashkent, Uzbekistan
e-mail: oqila1992@mail.ru

M. Ruzhansky, B. Torebek (eds.), *Extended Abstracts MWCAPDE 2023*,
Research Perspectives Ghent Analysis and PDE Center 1,
https://doi.org/10.1007/978-3-031-41665-1_13

1 Introduction

The main object of study in this paper is the following Rayleigh-Stokes problem for a generalized second-grade fluid with a time-fractional derivative model

$$\begin{cases} \partial_t u(x,t) - (1+\gamma\, \partial_t^\alpha)\Delta u(x,t) = f(x,t), & x \in \Omega, \quad 0 < t \le T; \\ u(x,t) = 0, \quad x \in \partial\Omega, \quad 0 < t \le T; \\ u(x,0) = \varphi(x), \quad x \in \Omega, \end{cases} \tag{1.1}$$

where $\gamma > 0$ is a fixed constant, f and φ are the given functions, $\partial_t = \partial/\partial t$, and ∂_t^α is the Riemann-Liouville fractional derivative of order $\alpha \in (0, 1)$ (see, e.g., [1], p. 69); $\Omega \subset \mathbb{R}^N$ is a domain with a smooth boundary $\partial\Omega$.

In recent years, the Rayleigh-Stokes problem (1.1) has received much attention due to its importance for applications (see, e.g., [2] - [15]). An overview of work in this direction can be found in Bazhlekova et al. [7] (see also [8]). Here, we only note the main directions in which research was carried out.

1. In order to get an idea of the behavior of the solution of this model, an exact solution has been obtained in some special cases; see, e.g., [4–6];
2. The Sobolev regularity of the homogeneous problem was studied in the fundamental work of Bazhlekov et al. [7] (for the case $f(x,t) \neq 0$, see [8]);
3. Several specialists have developed efficient and optimally accurate numerical algorithms to solve the problem (1.1). A review of some works in this direction is contained in the abovementioned paper [7]. See also recent papers [9, 10], and references therein;
4. Many works are devoted to the study of the inverse problem of determining the right-hand side of the Rayleigh-Stokes equation (see, e.g., [11–13], and the bibliography cited there). Since this inverse problem is Hadamard ill-posed, various regularization methods and numerical methods for finding the right-hand side of the equation are proposed in these works;
5. If the initial condition $u(x,0) = \varphi(x)$ in the problem (1.1) is replaced by $u(x,T) = \varphi(x)$, then the resulting problem is called *backward problem*. The backward problem for the Rayleigh-Stokes equation is of great importance and is aimed at determining the previous state of a physical field based on its current information (see, e.g., [14, 15] for the case $N \le 3$ and [8] for an arbitrary N). However, this problem is ill-posed according to Hadamard. Therefore, the authors of [14, 15] proposed various regularization methods and tested these methods using numerical experiments.

Let us focus, in more detail, on the recently published work [8]. In this paper, along with other questions, problem (1.1) is investigated by taking the non-local condition $u(x,T) = \beta u(x,0) + \varphi(x)$ instead of the initial condition. The authors considered only the cases $\beta = 0$ and $\beta = 1$: if $\beta = 0$ then we have the backward problem (note that here the dimension N is arbitrary). The authors proved that if $\beta = 0$, then the solution exists and is unique, but there is no stability, and if $\beta = 1$, then the problem is well-posed in the sense of Hadamard, i.e., the unique solution exists, and the solution continuously depends on the initial data and on the right-hand side of the equation.

The question naturally arises: what happens when β takes values other than 0 and 1? In Sect. 2 of the present paper, we consider a more general non-local condition $u(x, t_0) = \beta u(x, 0) + \varphi(x)$, $t_0 \in (0, T]$ and provide a definitive answer to this question.

It is well-known that the Rayleigh-Stokes problem (1.1) plays an important role in the study of the behavior of some non-Newtonian fluids. The fractional derivative ∂_t^α is used in Eq. (1.1) to describe the behavior of viscoelastic flow (see, e.g., [4, 5]). Obviously, knowing the exact value of this parameter is interesting not only theoretically; it is necessary for more adequate modeling of the physical process. But this parameter is often unknown and difficult to measure directly. Therefore, it is undoubtedly relevant to study the inverse problem to determine this a physical quantity from some indirectly observed information about the solutions. Such an inverse the problem would not only be of theoretical interest; it would help to more accurately determine the solutions of the initial-boundary-value problem and study the properties of the solutions.

Section 3 of this work is devoted to the study of this inverse problem for the Rayleigh-Stokes equation. To solve the inverse problem, it is necessary to set an additional condition. A natural requirement for such a condition is that it must ensure both the existence and the uniqueness of the unknown parameter. In this paper, it is shown that the additional condition $||u(x, t_0)||^2_{L_2(\Omega)} = d_0$ for a sufficiently large t_0 just satisfies these requirements.

Another very important problem in general for any fractional order equation is the study of the dependence of the behavior of the solution of the initial-boundary value problem on the order of the fractional derivative. To the best of our knowledge, this problem has not yet attracted sufficient attention from researchers. In this work, when solving the inverse problem, an interesting fact was discovered: if, for example, the norm $||u(x, t_0)||_{L_2(\Omega)}$, of the solution of (1.1), is considered as a function of the parameter α, then it is a decreasing function. In other words, the norm acquires its maximum value when the order of the fractional derivative is close to zero and its minimum value when this parameter is close to one.

2 Non-local Problem

Consider the following non-local problem for the Rayleigh-Stokes equation:

$$\begin{cases} \partial_t u(x,t) - (1+\gamma\, \partial_t^\alpha)\Delta u(x,t) = f(x,t), & x \in \Omega, \quad 0 < t \le T; \\ u(x,t) = 0, \quad x \in \partial\Omega, \quad 0 < t \le T; \\ u(x,t_0) = \beta u(x,0) + \varphi(x), \quad x \in \Omega, \quad t_0 \in (0,T], \end{cases} \tag{2.1}$$

where $\gamma > 0$ is a fixed constant, $\varphi \in L_2(\Omega)$, $f(\cdot, t) \in C((0, T]; L_2(\Omega))$, and β is an arbitrary real number. If $\beta = 0$, then this problem is called *the backward problem* (this case was studied in the works [14, 15] (here $N \leq 3$) and [8] (here N any natural number)).

Definition 2.1 A function $u(\cdot, t) \in C([0, T]; L_2(\Omega))$ is called **a solution** of the non-local Rayleigh-Stokes problem (2.1), if

$$\partial_t u(\cdot, t),\ \Delta u(\cdot, t),\ \partial_t^\alpha \Delta u(\cdot, t) \in C((0, T); L_2(\Omega)),$$

and it satisfies conditions (2.1) for all $t \in (0, T]$.

To solve the non-local problem (2.1), we use the method of separation of variables, i.e., the solution will be sought in the form of a series in terms of eigenfunctions of the spectral problem:

$$\begin{cases} -\Delta v(x) = \lambda v(x), & x \in \Omega; \\ v(x) = 0, & x \in \partial\Omega. \end{cases} \tag{2.2}$$

It is well-known that this problem has a complete system of eigenfunctions $\{v_k(x)\}$ orthonormal in $L_2(\Omega)$ and the set of positive eigenvalues $\{\lambda_k\}$: $0 < \lambda_1 < \lambda_2 \leq \lambda_3 \leq \cdots \to +\infty$ (see, e.g., [16]). The set $\{\lambda_k\}$ has no finite limit points. In particular, the multiplicity of any eigenvalue is finite.

After separating variables, we need to solve the following Cauchy problem for an ordinary differential equation:

$$Ly(t) \equiv y'(t) + \lambda(1 + \gamma\partial_t^\alpha)y(t) = 0,\ t > 0,\ \lambda > 0,\ y(0) = 1. \tag{2.3}$$

Let $B_\alpha(\lambda, t)$ be a solution of this problem. Function $B_\alpha(\lambda, t)$ is studied in detail in Bazhlekova, Jin, Lazarov, and Zhou [7]. The authors of [7], in particular, proved the following lemma.

Lemma 2.2 *Let $B_\alpha(\lambda, t)$ be a solution of the Cauchy problem (2.3). Then*

(1) $B_\alpha(\lambda, 0) = 1,\ 0 < B_\alpha(\lambda, t) < 1,\ t > 0,$
(2) $\lambda B_\alpha(\lambda, t) < C \min\{t^{-1}, t^{\alpha-1}\},\ t > 0.$

The function $B_\alpha(\lambda, t)$ has the representation [7]

$$B_\alpha(\lambda, t) = \int_0^\infty e^{-rt} b_\alpha(\lambda, r) dr, \tag{2.4}$$

where

$$b_\alpha(\lambda, r) = \frac{\gamma}{\pi} \frac{\lambda r^\alpha \sin \alpha\pi}{(-r + \lambda\gamma r^\alpha \cos \alpha\pi + \lambda)^2 + (\lambda\gamma r^\alpha \sin \alpha\pi)^2}.$$

When constructing a formal solution of the non-local problem by the Fourier method, we obtain a series whose denominators include the following expression $B_\alpha(\lambda_k, t_0) - \beta$. If $\beta \notin [0, 1)$, then taking into account the property of the function $B_\alpha(\lambda, t)$ (see Lemma 2.2), we can prove the convergence and differentiability of these series, i.e., make sure that in this case the solution exists and it is unique. Case $\beta = 0$ was considered in work [8].

Let $\beta \in (0, 1)$. Consider the following equation

$$B_\alpha(\lambda, t_0) = \beta \tag{2.5}$$

with respect to λ. The fact that a solution to this equation exists follows from Lemma (2.2). If $B_\alpha(\lambda, t_0)$ is strictly monotone as a function of λ, then the solution of the equation is unique. Therefore, the following lemmas are important for further research. From now on we assume that $\lambda \geq \lambda_1 > 0$.

Lemma 2.3 *Let* $0 < t_0 \leq T$, $\gamma > 0$, $\alpha \in (0, 1)$ *be given numbers. There exists a positive number* $\Lambda_0 = \Lambda_0(t_0, \gamma, \alpha, \lambda_1) > 0$ *such that for any* $\lambda \geq \Lambda_0$, *one has*

$$\partial_\lambda B_\alpha(\lambda, t_0) < 0. \tag{2.6}$$

If λ_0 is a root of Eq. (2.5), then the set of all k for which $\lambda_k = \lambda_0$ will be denoted by K_0. If there is no such eigenvalue λ_k, equal to λ_0, then naturally the set K_0 is empty.

Remark 2.4 According to Lemma 2.3, starting from some number k, the function $B_\alpha(\lambda_k, t_0)$ decreases strictly with respect to λ_k (if the multiplicity of the eigenvalue λ_k is not taken into account). Therefore, the set K_0 is always finite.

Thus, Lemma 2.3 states that only a finite number of eigenvalues λ_k can be solutions of Eq. (2.5). It can also be proved that if t_0 is large enough, then there can be only one such eigenvalue. Indeed, the following statement is true.

Lemma 2.5 *Let* $\gamma > 0$ *and* $\alpha \in (0, 1)$ *be given numbers. There exists a positive number* $T_0 = T_0(\gamma, \alpha, \lambda_1) \geq 1$ *such that for any* t_0 *in the interval* $T_0 \leq t_0 \leq T$ *one has*

$$\partial_\lambda B_\alpha(\lambda, t_0) < 0, \quad \lambda \geq \lambda_1. \tag{2.7}$$

Remark 2.6 Under the conditions of this lemma, there can be only one eigenvalue λ_{k_0} that satisfies Eq. (2.5). Let the multiplicity of λ_{k_0} be equal to p. Then $K_0 = \{k_0, k_0 + 1, \cdots, k_0 + p - 1\}$.

Let us pay attention to the fact that in Lemma 2.3, λ is a sufficiently large, and t_0 is an arbitrary positive number; in Lemma 2.5, on the contrary, t_0 is sufficiently large, and $\lambda \geq \lambda_1$ is an arbitrary number.

The proof of the next main result of this section is based on these formulated Lemmas 2.3 and 2.5.

Theorem 2.7 *Let* $\varphi \in L_2(\Omega)$ *and* $||f(\cdot, t) - f(\cdot, s)||_{L_2(\Omega)} \leq C_L|t-s|^{\varepsilon}$, $C_L > 0$, *for some* $\varepsilon \in (0, 1)$.

If $\beta \notin [0, 1)$ *or* $\beta \in (0, 1)$, *but* K_0 *is empty, then problem (2.1) has a unique solution, and this solution has the form*

$$u(x,t) = \sum_{k=1}^{\infty} \left[\frac{\varphi_k - \omega_k(t_0)}{B_\alpha(\lambda_k, t_0) - \beta} B_\alpha(\lambda_k, t) + \omega_k(t) \right] v_k(x),$$

where

$$\omega_k(t) = \int_0^t B_\alpha(\lambda_k, \eta) f_k(t - \eta) d\eta.$$

If $\beta \in (0, 1)$ *and* K_0 *is not empty, then then a solution to problem (2.1) exists if and only if the orthogonality conditions*

$$(\varphi, v_k) = (\omega(t_0), v_k), \ \ k \in K_0, \tag{2.8}$$

are satisfied. The solution of problem (2.1) is not unique, and all solutions have the form

$$u(x,t) = \sum_{k \notin K_0} \left[\frac{\varphi_k - \omega_k(t_0)}{B_\alpha(\lambda_k, t_0) - \beta} B_\alpha(\lambda_k, t) + \omega_k(t) \right] v_k(x) + \sum_{k \in K_0} h_k B_\alpha(\lambda_k, t) v_k(x),$$

with arbitrary coefficients h_k, $k \in K_0$.

Remark 2.8 The orthogonality condition (2.8) is difficult to verify, however, if the simpler condition

$$(\varphi, v_k) = 0, \ \ (f(t), v_k) = 0, \ \ \text{for all } 0 \leq t \leq t_0, \ \ k \in K_0,$$

holds, then (2.8)) also holds.

3 Inverse Problem

Consider the homogeneous Rayleigh-Stokes problem

$$\begin{cases} \partial_t u(x,t) - (1+\gamma\, \partial_t^\alpha)\Delta u(x,t) = 0, \quad x \in \Omega, \quad 0 < t \le T; \\ u(x,t) = 0, \quad x \in \partial\Omega, \quad 0 < t \le T; \\ u(x,0) = \varphi(x), \quad x \in \Omega, \end{cases} \tag{3.1}$$

where $\gamma > 0$, $\alpha \in (0,1)$ and $\varphi \in L_2(\Omega)$.

Assume that the order α of the fractional derivative in this problem is unknown. Consider the inverse problem of determining this parameter. The homogeneity of the equation greatly facilitates the application of the method proposed in this paper.

To find α, we propose the following additional condition:

$$U(t_0,\alpha) \equiv ||u(\cdot,t_0)||^2_{L_2(\Omega)} = d_0, \quad t_0 \ge T_0, \tag{3.2}$$

where T_0 is defined later. Problem (3.1) together with the additional condition (3.2) will be called *the inverse problem*.

Before stating the main result of this section on the solution of the inverse problem, we formulate an auxiliary assertion, which is also of independent interest.

Theorem 3.1 *There exists a number $T_0 = T_0(\gamma,\alpha,\lambda_1) \ge 1$ such that for every $t_0 \ge T_0$ and every $\varphi \in L_2(\Omega)$ the function $U(t_0,\alpha)$ strictly decreasing in $\alpha \in (0,1)$.*

This theorem establishes the dependence of the norm of the solution of the Rayleigh-Stokes problem on the order of the fractional derivative. There are only a few works where such a dependence of the solution of initial-boundary value problems for subdiffusion equations has been studied (see, e.g., [17] and [18]).

Here is the main result of this section.

Theorem 3.2 *Let $\varphi \in L_2(\Omega)$ and $t_0 \ge T_0$. Then for the inverse problem to have a solution $\{u(x,t),\alpha\}$, it is necessary and sufficient that the condition*

$$\inf_{\alpha\in(0,1)} U(t_0,\alpha) \le d_0 \le \sup_{\alpha\in(0,1)} U(t_0,\alpha) \tag{3.3}$$

be satisfied.

Remark 3.3 Theorem defines the unique α from (3.2). Hence, if we define the norm $U(t_0,\alpha)$ at another time instant $t_1 \ge T_0$ and get a new α_1, i.e., $U(t_1,\alpha_1) = d_1$, then from the equality $U(t_0,\alpha_1) = d_0$, by virtue of the theorem, we obtain $\alpha_1 = \alpha$.

The inverse problem of determining the order of the fractional derivative for the subdiffusion equations, the fractional-wave equation, and mixed-type equations has been studied by many authors. An overview of works published up to 2019 can be found in [19]. After this paper, a lot of works have been devoted to this subject, and a brief review of these works can be found in works [20] and [21].

Proofs of these theorems are based on the following lemma.

Lemma 3.4 *Let* $\gamma > 0$, $\lambda \geq \lambda_1$ *and* $\alpha \in (0, 1)$ *be given numbers. There exists a positive number* $T_0 = T_0(\gamma, \alpha, \lambda_1) \geq 1$ *such that for any* t_0, $T_0 \leq t_0 \leq T$, *one has*

$$\partial_\alpha B_\alpha(\lambda, t_0) < 0. \tag{3.4}$$

To prove this lemma, we first make a change of variable $rt_0 = \xi$ in the integral for $B_\alpha(\lambda, t_0)$. Then

$$B_\alpha(\lambda, t_0) = \frac{1}{t_0}\int_0^\infty e^{-\xi}\, b_\alpha\left(\lambda, \frac{1}{t_0}\xi\right) d\xi = t_0^{\alpha-1}\int_0^\infty e^{-\xi} b_{\alpha,1}(\lambda, t_0, \xi) d\xi, \tag{3.5}$$

where the function $b_{\alpha,1}(\lambda, t_0, \xi)$ is easily calculated. Differentiating with respect to α, we show that the main part of the resulting expression is that part of the derivative that contains the factor $t_0^{\alpha-1}\ln t_0$. Then we show that the principal part is negative for sufficiently large t_0.

If parameter α is known, then the Rayleigh-Stokes problem (3.1) is solved, for example, in work [7] (see also [8]): for any $\varphi \in L_2(\Omega)$, the problem has a unique solution

$$u(x, t) = \sum_{k=1}^\infty B_\alpha(\lambda_k, t)\varphi_k v_k(x).$$

Therefore,

$$U(t, \alpha) = ||u(\cdot, t)||^2_{L_2(\Omega)} = \sum_{k=1}^\infty B_\alpha^2(\lambda_k, t)|\varphi_k|^2.$$

Lemma 3.4 proves that the derivative $\partial_\alpha B_\alpha(\lambda_k, t_0)$ is negative for a sufficiently large t_0 and for all $k \geq 1$, i.e., it is proved that for all $k \geq 1$ and starting from $t_0 \geq T_0$ function, $B_\alpha(\lambda_k, t_0)$ decreases with respect to α. The proof of Theorem 3.1 is based on this property of the derivative $\partial_\alpha B_\alpha(\lambda_k, t_0)$ and on the fact that according to Lemma 2.2, $B_\alpha(\lambda_k, t) > 0$ for all $t > 0$ and $k \geq 1$.

Regarding the proof of Theorem 3.2, we note the following. By the conditions of the theorem (see (3.3), let the given number d_0 be such that

$$\inf_{\alpha\in(0,1)} U(t_0, \alpha) \leq d_0 \leq \sup_{\alpha\in(0,1)} U(t_0, \alpha), \quad t_0 \geq T_0.$$

Considering that, according to Theorem 3.1, the function $U(t_0, \alpha)$ is strictly monotonic, this immediately implies the existence of a single number α satisfying the additional condition (3.2). Obviously, if the opposite inequalities hold, then the number d_0 does not enter the range of the function $U(t_0, \alpha)$, and as a result, the number α satisfying the additional condition (3.2) does not exist.

After the parameter α is found, the solution $u(x, t)$ that satisfies all the requirements of Definition 2.1 is found in the same way as in work [8].

To prove the uniqueness of the solution of the inverse problem, we assume the existence of two pairs of solutions $\{u_1, \alpha_1\}$ and $\{u_2, \alpha_2\}$. Note that the inverse problem is nonlinear with respect to u and α. For $t_0 \geq T_0$ additional condition (3.2) implies

$$||u_1(\cdot, t_0)||^2_{L_2(\Omega)} = ||u_2(\cdot, t_0)||^2_{L_2(\Omega)} = d_0. \tag{3.6}$$

Assume, for example, that $\alpha_1 < \alpha_2$ and $t_0 \geq T_0$. Then by Lemma 3.4

$$|\varphi_k|^2 B^2_{\alpha_1}(\lambda_k, t_0) > |\varphi_k|^2 B^2_{\alpha_2}(\lambda_k, t_0), \quad k \geq 1.$$

Therefore $||u_1(\cdot, t_0)||^2_{L_2(\Omega)} > ||u_2\cdot, t_0)||^2_{L_2(\Omega)}$ and this contradicts equality (3.6). Consequently, $\alpha_1 = \alpha_2$ and further it is easy to prove that $u_1(x, t) \equiv u_2(x, t)$.

Remark 3.5 The function $U(t_0, \alpha)$ in the additional condition (3.2) can also be taken in the form $||\Delta u(\cdot, t_0)||^2_{L_2(\Omega)}$ or $(\Delta u(\cdot, t_0), u(\cdot, t_0))$ (the inner product in $L_2(\Omega)$).

Remark 3.6 Instead of the Laplace operator with the Dirichlet condition, we can take the operators of mathematical physics considered in Section 6 of the paper by M. Ruzhansky et al. [22], including the classical Sturm-Liouville problem, differential models with involution, fractional Sturm-Liouville operators, harmonic and anharmonic oscillators, Hamiltonians Landau, fractional Laplacians, and harmonic and anharmonic operators on the Heisenberg group. As a result, we obtain various Rayleigh-Stokes equations for which the formulated above theorems will be valid.

Acknowledgments The first author acknowledges financial support from the Ministry of Innovative Development of the Republic of Uzbekistan, Grant No F-FA-2021–424.

References

1. A. Kilbas, H. Srivastava, J. Trujillo, *Theory and Applications of Fractional Differential Equations* (Elsevier, Amsterdam, 2006)
2. W.C. Tan, T. Masuoka, Stokes' first problem for a second grade fluid in a porous half-space with heated boundary. Int. J. Non-Linear Mech. **40**, 515–522 (2005)
3. W.C. Tan, T. Masuoka, Stokes' first problem for an Oldroyd-B fluid in a porous half-space. Phys. Fluid **17**, 023101-7 (2005)
4. C. Fetecau, M. Jamil, C. Fetecau, D. Vieru, The Rayleigh-Stokes problem for an edge in a generalized Oldroyd-B fluid. Z. Angew. Math. Phys. **60**(5), 921–933 (2009)

5. F. Shen, W. Tan, Y. Zhao, T. Masuoka, The Rayleigh-Stokes problem for a heated generalized second grade fluid with fractional derivative model. Nonlinear Anal. Real World Appl. **7**(5), 1072–1080 (2006)
6. C. Zhao, C. Yang, Exact solutions for electro-osmotic flow of viscoelastic fluids in rectangular micro-channels. Appl. Math. Comput. **211**(2), 502–509 (2009)
7. E. Bazhlekova, B. Jin, R. Lazarov, Z. Zhou, An analysis of the Rayleigh–Stokes problem for a generalized second-grade fluid. Numer. Math. **131**, 1–31 (2015)
8. R. Ashurov, N. Vaisova, Backward and non-local problems for the Rayleigh-Stokes equation. Fractal Fract. **6**(10), 587 (2022)
9. L.D. Long, B. Moradi, O. Nikan, Z. Avazzadeh, A.M. Lopes, Numerical approximation of the fractional Rayleigh-Stokes problem arising in a generalised Maxwell fluid. J. Fractal Fract. **6**, 377 (2022)
10. D.D. Dai, T.T. Ban, Y.L. Wang, W. Zhang, T. Hang, The piecewise reproducing kernel method for the time variable fractional order advection-reaction-diffusion equations. Therm. Sci. **25**, 1261–1268 (2021)
11. T.T. Binh, D. Baleanu, N.H. Luc, N.-H. Can, Determination of source term for the fractional Rayleigh–Stokes equation with random data. J. Inequal. Appl. **308**, 1–16 (2019)
12. T.T. Binh, H.K. Nashine, L.D. Long, N.H. Luc, C. Nguyen, Identification of source term for the ill-posed Rayleigh–Stokes problem by Tikhonov regularization method. Adv. Differ. Equ. **331**, 1–20 (2019)
13. P.N. Duc, H.D. Binh, L.D. Long, H.T.K. Van, Reconstructing the right-hand side of the Rayleigh–Stokes problem with nonlocal in time condition. Adv. Differ. Equ. **470**, 1–18 (2021)
14. H.L. Nguyen, H.T. Nguyen, K. Mokhtar, X.T.D. Dang, *Identifying Initial Condition of the Rayleigh-Stokes Problem with Random Noise* Math. Methods Appl. Sci. **42**, 1561–1571 (2019)
15. H.L. Nguyen, L.N. Huynh, D. O'Regan, N.H. Can, Regularization of the fractional Rayleigh–Stokes equation using a fractional Landweber method. Adv. Differ. Equ. **459**, 1–23 (2020)
16. Y.M. Berezanskii, *Expansions in Eigenfunctions of Selfadjoint Operators*, Translations of Mathematical Monographs, vol. 17 (American Mathematical Society, Providence, RI, 1968)
17. S. Alimov, R. Ashurov, Inverse problem of determining an order of the Caputo time-fractional derivative for a subdiffusion equation. J. Inverse Ill-Posed Probl. **28**, 651–658 (2020)
18. S. Alimov, R. Ashurov, Inverse problem of determining an order of the Riemann-Liouville time-fractional derivative. Progr. Fract. Differ. Appl. **8**, 1–8 (2022)
19. Z. Li, Y. Liu, M. Yamamoto, Inverse problems of determining parameters of the fractional partial differential equations, in *Handbook of Fractional Calculus with Applications*, vol. 2 (De Gruyter, Berlin, Germany, 2019)
20. R.R. Ashurov, Y.E. Fayziev, Inverse problem for determining the order of a fractional derivative in a wave equation. Math. Notes **110**(6), 824–836 (2021)
21. R.R. Ashurov, S. Sitnik, Identification of the order of the fractional derivative for the fractional wave equation. Fractal Fract. **7**(1), 67 (2023)
22. M. Ruzhansky, N. Tokmagambetov, B.T. Torebek, Inverse source problems for positive operators. I. Hypoelliptic diffusion and subdiffusion equations. J. Inverse Ill-Possed Probl. **27**, 891–911 (2019)

Very Weak Solution of the Discrete Heat Equation with Irregular Time-Dependent Thermal Conductivity

Marianna Chatzakou and Abhilash Tushir

Abstract In this paper, we investigate the semi-classical version of the heat equation with irregular time-dependent coefficients. When regular coefficients are taken into account, we show that the Cauchy problem is well-posed in $\ell^2(\hbar\mathbb{Z}^n)$. In the case of irregular coefficients, we analyse how the notion of a very weak solution adapts to our consideration and show that the Cauchy problem is "weakly well-possed" and that the very weak solution converges approximates (in a suitable norm sense) the classical solution. This presentation is based on the paper Chatzakou et al. (Proc R Soc Edinb A: Math, 1–24, 2023. https://doi.org/10.1017/prm.2023.84), to which we refer for further details.

1 Introduction

The heat equation with constant, time-dependent, or space-dependent thermal coefficients has been extensively considered in a wide range of settings. For the case of the heat equation with space-dependent singular potentials, we refer to the works [1, 5]. For the case where the time-dependent coefficients are considered, we refer to the works [10–12] and [15]. In the particular case of the lattice $\mathbb{Z}^n$, we have the work [8] where constant coefficients are considered. In [6, 7] the authors examine the discrete wave equation with time-dependent coefficients and the discrete Klein-Gordon equation with regular potential and prove that they are well-posed in $\ell^2(\hbar\mathbb{Z}^n)$.

M. Chatzakou
Department of Mathematics: Analysis, Logic and Discrete Mathematics, Ghent University, Ghent, Belgium
e-mail: marianna.chatzakou@ugent.be

A. Tushir (✉)
Department of Mathematics, Indian Institute of Technology, Delhi, New Delhi, India

M. Ruzhansky, B. Torebek (eds.), *Extended Abstracts MWCAPDE 2023*,
Research Perspectives Ghent Analysis and PDE Center 1,
https://doi.org/10.1007/978-3-031-41665-1_14

In this work, the space variable lies in the setting of the discrete lattice

$$\hbar\mathbb{Z}^n = \{x \in \mathbb{R}^n : x = \hbar k,\ k \in \mathbb{Z}^n\},$$

depending on a (small) semi-classical parameter $\hbar > 0$.

The heat equation in this setting for $a = a(t) \geq 0$ being the thermal conductivity, b being a real-valued bounded potential, and f being the heat source reads as follows:

$$\begin{cases} \partial_t u(t,k) - a(t)\hbar^{-2}\mathcal{L}_\hbar u(t,k) + b(t)u(t,k) = f(t,k), & (t,k) \in (0,T] \times \hbar\mathbb{Z}^n, \\ u(0,k) = u_0(k), \quad k \in \hbar\mathbb{Z}^n, \end{cases} \tag{1.1}$$

which is the semi-classical discrete analog of the heat equation in the Euclidean setting given by the Cauchy problem

$$\begin{cases} \partial_t u(t,x) - a(t)\mathcal{L}u(t,x) + b(t)u(t,x) = f(t,x), & (t,x) \in (0,T] \times \mathbb{R}^n, \\ u(0,x) = u_0(x), \quad x \in \mathbb{R}^n, \end{cases}$$

for $\mathcal{L}$ being the usual Laplacian on $\mathbb{R}^n$.

We note that the discrete Laplacian $\mathcal{L}_\hbar$ on the lattice $\hbar\mathbb{Z}^n$ as appears in (1.1) is defined as

$$\mathcal{L}_\hbar u(k) = \sum_{l=1}^{n} (u(k + v_l\hbar) + u(k - v_l\hbar)) - 2nu(k),$$

where v_l is the lth basis vector in $\mathbb{Z}^n$, having all zeros except for 1 at the lth component.

The Cauchy problem (1.1) is a particular case of the one analysed in [3] where the authors allow fractional powers of the discrete Laplacian $\mathcal{L}_\hbar$ and for the source term one has $f \in L^2([0,T]; \ell^2_s(\hbar\mathbb{Z}^n))$ for any real $s \in \mathbb{R}$. Note that the space $\ell^2(\hbar\mathbb{Z}^n)$ is a space case of $\ell^2_s(\hbar\mathbb{Z}^n)$ for $s = 0$. Particularly, in the current work, we announce the results of the work [3] for $f \in L^2([0,T]; \ell^2(\hbar\mathbb{Z}^n))$ and without allowing for fractional powers of $\mathcal{L}_\hbar$. For a more detailed analysis and for the proofs of the announced results, we refer to [3].

A special characteristic of our analysis is that the coefficients a, b can have distributional irregularities. This consideration would bring fundamental mathematical difficulties in the sense that since the solution $u(t,x)$ might as well have singularities in t; see Schwartz's result [14] on the impossibility of multiplying distributions. To overcome this issue we make use of the theory of very weak solutions introduced in [9] that allows for the aforementioned consideration. Notably the approach we develop allows recovering the classical solution to the problem (1.1).

2 Preliminaries

In [13] and [2], the authors developed the Fourier analysis related to the discrete lattice $\mathbb{Z}^n$. In this section, we summarise the required preliminaries for our analysis as appeared in these works.

The Fourier transform operator

$$\mathcal{F}_{\hbar\mathbb{Z}^n} : \ell^2(\hbar\mathbb{Z}^n) \to L^2(\mathbb{T}^n)$$

is defined as

$$\mathcal{F}_{\hbar\mathbb{Z}^n} u(\xi) := \sum_{k\in\hbar\mathbb{Z}^n} u(k)e^{-2\pi i \frac{k}{\hbar}\cdot\xi}, \quad \xi \in \mathbb{T}^n.$$

For the inverse Fourier transform operator

$$\mathcal{F}^{-1}_{\hbar\mathbb{Z}^n} : L^2(\mathbb{T}^n) \to \ell^2(\hbar\mathbb{Z}^n)$$

we have

$$\mathcal{F}^{-1}_{\hbar\mathbb{Z}^n} v(k) := \int_{\mathbb{T}^n} v(\xi)e^{2\pi i \frac{k}{\hbar}\cdot\xi} \mathrm{d}\xi, \quad k \in \hbar\mathbb{Z}^n,$$

while the Fourier inversion formula is given by

$$u(k) = \int_{\mathbb{T}^n} \widehat{u}(\xi)e^{2\pi i \frac{k}{\hbar}\cdot\xi} \mathrm{d}\xi, \quad k \in \hbar\mathbb{Z}^n.$$

The Fourier transform of the discrete Laplacian $\mathcal{L}_\hbar$ can be calculated using the shifting property of the Fourier transform as follows:

$$\mathcal{F}_{\hbar\mathbb{Z}^n} \mathcal{L}_\hbar u(\xi) = \sum_{k\in\hbar\mathbb{Z}^n} \mathcal{L}_\hbar u(k)e^{-2\pi i \frac{k}{\hbar}\cdot\xi} = -\sum_{j=1}^{n} 4\sin^2\left(\pi\xi_j\right)\widehat{u}(\xi)$$

for all $\xi \in \mathbb{T}^n$. For more details about the discrete Laplacian on $\hbar\mathbb{Z}^n$, see [6].

Finally, as for the regularisation of the distributional coefficients, this must be regarded as follows: let $a \in \mathcal{D}'(\mathbb{R})$ be a distribution. Using the Friedrichs-mollifier, i.e. a function $\psi \in C_0^\infty(\mathbb{R})$, $\psi \geq 0$ such that $\int_\mathbb{R} \psi = 1$, one can construct families of smooth functions $(a_\varepsilon)_\varepsilon$ via the convolution

$$a_\varepsilon(t) := \left(a * \psi_{\omega(\varepsilon)}\right)(t),$$

where $\psi_{\omega(\varepsilon)}$ is defined as

$$\psi_{\omega(\varepsilon)}(t) = \frac{1}{\omega(\varepsilon)}\psi\left(\frac{t}{\omega(\varepsilon)}\right),$$

for some $\omega(\varepsilon) \geq 0$ such that $\omega(\varepsilon) \to 0$ as $\varepsilon \to 0$.

Finally, let us note that in the sequel when we say that distribution a is positive, we mean that $\langle a, \psi\rangle \geq 0$ for all $\psi \in C^\infty(\mathbb{R})$ satisfying $\psi \geq 0$. Similarly, a distribution a shall be strictly positive whenever there exists a positive constant α such that $a - \alpha$ is a positive distribution.

3 Main Results

In this section, we state our results on the well-posedness of the Cauchy problem (1.1) with regular and irregular coefficients. For the latter case, the notion of a very weak solution and its uniqueness to the Cauchy problem (1.1) is required. Our last result is on the consistency between the very weak solution and the classical one.

The well-posedness theorem for the Cauchy problem (1.1) with regular coefficients is as follows.

Theorem 3.1 (Classical solution) *Let* $f \in L^2([0, T]; \ell^2(\hbar\mathbb{Z}^n))$. *Assume that* $a \in L_1^\infty([0, T])$[1] *satisfies* $\inf_{t\in[0,T]} a(t) = a_0 > 0$ *and* $b \in L^\infty([0, T])$. *If* $u_0 \in \ell^2(\hbar\mathbb{Z}^n)$, *then the Cauchy problem* (1.1) *has a unique solution* $u \in C([0, T]; \ell^2(\hbar\mathbb{Z}^n))$ *satisfying the estimate*

$$\|u(t, \cdot)\|^2_{\ell^2(\hbar\mathbb{Z}^n)} \lesssim \|u_0\|^2_{\ell^2(\hbar\mathbb{Z}^n)} + \|f\|^2_{L^2([0,T];\ell^2(\hbar\mathbb{Z}^n))},^2 \tag{3.1}$$

for all $t \in [0, T]$.

In the next theorem, we allow the coefficients a, b as well as the source term f to have singularities in the time variable. This consideration requires introducing the notion of a very weak solution for our Cauchy problem (1.1). The notion of very weak solution involves in turn the notions of moderateness and negligibility of a net of functions/distributions which is as follows.

[1] We write $a \in L_1^\infty([0, T])$, whenever $a \in L^\infty([0, T])$ and $\partial_t a \in L^\infty([0, T])$.

[2] We write $a \lesssim b$ whenever there exists a constant C, independent of a and b, such that $a \leq Cb$.

Definition 3.2

(i) A net of functions $(a_\varepsilon)_\varepsilon \in X$ is said to be X-moderate if there exist $N \in \mathbb{N}_0$ and $c > 0$ such that

$$\|a_\varepsilon\|_X \leq c\varepsilon^{-N},$$

for all $\varepsilon \in (0, 1]$.

(ii) A net of functions $(a_\varepsilon)_\varepsilon \in X$ is said to be X-negligible if there exists $c > 0$ such that

$$\|a_\varepsilon\|_X \leq c\varepsilon^q, \qquad \text{for all } q \in \mathbb{N}_0,$$

for all $\varepsilon \in (0, 1]$.

(iii) A net of functions $(u_\varepsilon)_\varepsilon \in L^2([0, T]; X)^{(0,1]}$ is said to be $L^2([0, T]; X)$-moderate if there exist $N \in \mathbb{N}_0$ and $c > 0$ such that

$$\|u_\varepsilon\|_{L^2([0,T];X)} \leq c\varepsilon^{-N},$$

for all $\varepsilon \in (0, 1]$.

(iv) A net $(u_\varepsilon)_\varepsilon \in L^2([0, T]; X)^{(0,1]}$ is said to be $L^2([0, T]; X)$-negligible if for all $q \in \mathbb{N}_0$ there exists $c > 0$ such that

$$\|u_\varepsilon\|_{L^2([0,T];X)} \leq c\varepsilon^q,$$

for all $\varepsilon \in (0, 1]$.

For the regularisation of the distributions a, b, f as in (1.1), the regularised version of (1.1) becomes

$$\begin{cases} \partial_t u_\varepsilon(t, k) - a_\varepsilon(t)\hbar^{-2}\mathcal{L}_\hbar u_\varepsilon(t, k) + b_\varepsilon(t)u_\varepsilon(t, k) = f_\varepsilon(t, k), & t \in (0, T], \\ u_\varepsilon(0, k) = u_0(k), \quad k \in \hbar\mathbb{Z}^n. \end{cases} \tag{3.2}$$

Definition 3.3 (Very weak solution) Let the coefficients a, b and the source term $f(\cdot, k)$ in (1.1) be the distributions for all $k \in \hbar\mathbb{Z}^n$. If there exist

- a L_1^∞-moderate regularisation of the coefficient a;
- a L^∞-moderate regularisation of the coefficient b;
- a $L^2([0, T]; \ell^2(\hbar\mathbb{Z}^n))$-moderate regularisation of the source term f; and
- the solution to the Cauchy problem (3.2) is $L^2([0, T]; \ell^2(\hbar\mathbb{Z}^n))$-moderate,

then, the net $(u_\varepsilon)_\varepsilon$ is said to be a very weak solution to the Cauchy problem (1.1).

The existence theorem for the Cauchy problem (1.1) with distributional coefficients is as follows.

Theorem 3.4 (Existence) *Let a and b be distributions with supports contained in $[0, T]$ such that $a \geq a_0 > 0$ for some positive constant a_0, and let the source term $f(\cdot, k)$ be a distribution with support contained in $[0, T]$, for all $k \in \hbar\mathbb{Z}^n$. If for the initial Cauchy data we have $u_0 \in \ell^2(\hbar\mathbb{Z}^n)$, then the Cauchy problem (1.1) has a very weak solution.*

We note that the uniqueness of the very weak solution can be interpreted as if "negligible modifications" of the approximations of the coefficients a, b and the source term f result into negligible changes of the very weak solution. The notion of uniqueness can be formulated following the lines of [4, 5].

Before giving a formal definition of the unique very weak solution to our Cauchy problem (1.1), let us first explicitly write two regularised versions of it:

$$\begin{cases} \partial_t u_\varepsilon(t,k) - a_\varepsilon(t)\hbar^{-2}\mathcal{L}_\hbar u_\varepsilon(t,k) + b_\varepsilon(t)u_\varepsilon(t,k) = f_\varepsilon(t,k), & t \in (0,T], \\ u_\varepsilon(0,k) = u_0(k), \quad k \in \hbar\mathbb{Z}^n, \end{cases} \tag{3.3}$$

and

$$\begin{cases} \partial_t \tilde{u}_\varepsilon(t,k) - \tilde{a}_\varepsilon(t)\hbar^{-2}\mathcal{L}_\hbar \tilde{u}_\varepsilon(t,k) + \tilde{b}_\varepsilon(t)\tilde{u}_\varepsilon(t,k) = \tilde{f}_\varepsilon(t,k), & t \in (0,T], \\ \tilde{u}_\varepsilon(0,k) = u_0(k), \quad k \in \hbar\mathbb{Z}^n. \end{cases} \tag{3.4}$$

Definition 3.5 (Unique very weak solution) Let the coefficients a, b and the source term $f(\cdot, k)$ in (1.1) be the distributions for all $k \in \hbar\mathbb{Z}^n$. The Cauchy problem (1.1) is said to have a $L^2([0, T]; \ell^2(\hbar\mathbb{Z}^n))$-unique very weak solution, if

- for all L_1^∞-moderate nets $a_\varepsilon, \tilde{a}_\varepsilon$ such that $(a_\varepsilon - \tilde{a}_\varepsilon)_\varepsilon$ is L_1^∞-negligible;
- for all L^∞-moderate nets $b_\varepsilon, \tilde{b}_\varepsilon$ such that $(b_\varepsilon - \tilde{b}_\varepsilon)_\varepsilon$ are L^∞-negligible; and
- for all $L^2([0, T]; \ell^2(\hbar\mathbb{Z}^n))$-moderate nets $f_\varepsilon, \tilde{f}_\varepsilon$ such that $(f_\varepsilon - \tilde{f}_\varepsilon)_\varepsilon$ is $L^2([0, T]; \ell^2(\hbar\mathbb{Z}^n))$-negligible,

the net $(u_\varepsilon - \tilde{u}_\varepsilon)_\varepsilon$ is $L^2([0, T]; \ell^2(\hbar\mathbb{Z}^n))$-negligible, where u_ε and $\tilde{u}_\varepsilon$ are the families of solutions corresponding to the ε-parametrised problems (3.3) and (3.4), respectively.

The following theorem gives the uniqueness of the very weak solution to the Cauchy problem (1.1) in the sense of Definition 3.5.

Theorem 3.6 (Uniqueness) *Let a and b be distributions with supports contained in $[0, T]$ such that $a \geq a_0 > 0$ for some positive constant a_0, and let the source term $f(\cdot, k)$ be a distribution with support contained in $[0, T]$, for all $k \in \hbar\mathbb{Z}^n$. If $u_0 \in \ell^2(\hbar\mathbb{Z}^n)$, then the very weak solution of the Cauchy problem (1.1) is $L^2([0, T]; \ell^2(\hbar\mathbb{Z}^n))$-unique.*

The next theorem states that the very weak solution obtained in Theorem 3.4 recovers the classical solution obtained in Theorem 3.1.

Theorem 3.7 (Consistency) *Let* $f \in L^2([0,T], \ell^2(\hbar\mathbb{Z}^n))$. *Assume that* $a \in L_1^\infty([0,T])$ *satisfies* $\inf_{t\in[0,T]} a(t) = a_0 > 0$ *and* $b \in L^\infty([0,T])$. *If the initial Cauchy data* $u_0 \in \ell^2(\hbar\mathbb{Z}^n)$, *then the regularised net* u_ε *converges to the classical solution of the Cauchy problem* (1.1), *in* $L^2([0,T]; \ell^2(\hbar\mathbb{Z}^n))$ *as* $\varepsilon \to 0$.

Acknowledgments We would like to thank Prof. Michael Ruzhansky and Prof. Aparajita Dasgupta for their insightful reviews and suggestions. M. Chatzakou is a postdoctoral fellow of the Research Foundation – Flanders (FWO) under the postdoctoral grant No 12B1223N. A.Tushir is supported by the institute assistantship from Indian Institute of Technology Delhi, India.

References

1. A. Altybay, M. Ruzhansky, M.E. Sebih, N. Tokmagambetov, The heat equation with strongly singular potentials. Appl. Math. Comput. **399**, 15 (2021). Paper No. 126006
2. L.N.A. Botchway, P. Gaël Kibiti, M. Ruzhansky, Difference equations and pseudo-differential operators on $\mathbb{Z}^n$. J. Funct. Anal. **278**(11), 108473, (2020). 41
3. M. Chatzakou, A. Dasgupta, M. Ruzhansky, A. Tushir, (2023). Discrete heat equation with irregular thermal conductivity and tempered distributional data. Proceedings of the Royal Society of Edinburgh Section A: Mathematics, 1–24. https://doi.org/10.1017/prm.2023.84
4. M. Chatzakou, M. Ruzhansky, N. Tokmagambetov, Fractional Klein-Gordon equation with singular mass. II: hypoelliptic case. Complex Var. Elliptic Equ. **67**(3), 615–632 (2022)
5. M. Chatzakou, M. Ruzhansky, N. Tokmagambetov, The heat equation with singular potentials. II: hypoelliptic case. Acta Appl. Math. **179**, 20 (2022). Paper No. 2
6. A. Dasgupta, M. Ruzhansky, A. Tushir, Discrete time-dependent wave equations I. Semiclassical analysis. J. Differ. Equ. **317**, 89–120 (2022)
7. A. Dasgupta, M. Ruzhansky, A. Tushir, Discrete time-dependent wave equations II. Semiclassical fractional Klein-Gordon equation (2022). arxiv: 2205.05360
8. R. Floreanini, J. Negro, L.M. Nieto, L. Vinet, Symmetries of the heat equation on the lattice. Lett. Math. Phys. (4) **36**, 351–355 (1996)
9. C. Garetto, M. Ruzhansky, Hyperbolic second order equations with non-regular time dependent coefficients. Arch. Ration. Mech. Anal. **217**(1), 113–154 (2015)
10. A. Gulisashvili, On the heat equation with a time-dependent singular potential. J. Funct. Anal. **194**(1), 17–52 (2002)
11. M.S. Hussein, D. Lesnic, M.I. Ivanchov, Simultaneous determination of time-dependent coefficients in the heat equation. Comput. Math. Appl. **67**(5), 1065–1091 (2014)
12. M. Ljulj, K. Schmidt, A. Semin, J. Tambača, Homogenization of the time-dependent heat equation on planar one-dimensional periodic structures. Appl. Anal. **101**(12), 4046–4075 (2022)
13. M. Ruzhansky, V. Turunen, Pseudo-differential operators and symmetries. Background analysis and advanced topics, in *Pseudo-Differential Operators. Theory and Applications*, vol. 2 (Birkhäuser, Basel, 2010)
14. L. Schwartz, Sur l'impossibilité de la multiplication des distributions. C. R. Acad. Sci. Paris **239**, 847–848 (1954)
15. J. Takahashi, E. Yanagida, Time-dependent singularities in the heat equation. Commun. Pure Appl. Anal. **14**(3), 969–979 (2015)

On a Mixed Stokes Problem with Variable Viscosity in the 2D Exterior Domain

Mulugeta A. Dagnaw

Abstract The boundary-domain integral equations for the mixed boundary value problem for a compressible Stokes system of partial differential equations with variable viscosity in the two-dimensional unbounded domain are considered. An appropriate parametrix is used to reduce this problem to some direct segregated boundary-domain integral equations (BDIEs). The equivalence of the original BVP and the obtained BDIEs are analysed in weighted Sobolev spaces.

2000 Mathematics Subject Classification 76N06, 76N20, 31A10,76N20

1 Introduction

The Stokes system of PDE is derived from the linearised steady-state Navier-Stokes system. Studying this system gives us an opportunity to introduce several tools necessary for a treatment of the full Navier-Stokes equations (see [17]). The Stokes system that we consider in this dissertation is elliptic in the sense of Douglis-Nirenberg [10].

It is well-known that boundary integral equations (BIEs) are one of the solution techniques for such problems that can be found by the reduction of boundary value problems (BVPs) for PDEs with constant coefficients. A more detailed description of BIEs can be found in [10, 11, 16].

The BIE method is one of the most powerful techniques developed by analysts for solving elliptic BVPs with constant coefficients. The advantage of this method is that it reduces a BVP for linear PDE in a domain to an equivalent integral equation on the boundary of the domain. Therefore, it diminishes the problem dimensionality by one, which is very important for the construction of various numerical algorithms [6].

M. A. Dagnaw (✉)
Injibara University, Injibara, Ethiopia

M. Ruzhansky, B. Torebek (eds.), *Extended Abstracts MWCAPDE 2023*,
Research Perspectives Ghent Analysis and PDE Center 1,
https://doi.org/10.1007/978-3-031-41665-1_15

The main ingredient for reducing a BVP for a PDE to a BIE is a fundamental solution to the original PDE. However, the reduction of a BVP with arbitrarily variable coefficients to boundary integral equations is usually not effective for numerical implementations. This is due to the fact that the fundamental solution which is necessary for such reduction is generally not available in an analytical form. In this case, following E. E. Levi and D.Hilbert (see [14, 15]), we can use parametrix (Levi function) to the original PDE as a substitute for the fundamental solution. This is because parametrix is usually much more largely available than the fundamental solution and correctly describes the main part of the fundamental solution. Note that the parametrix does not have to satisfy the original PDE. This reduces the problems to *boundary-domain integral equation(BDIE)* not to BIE [6].

Since the last decade, the theory of direct segregated and united BDIE systems for scalar PDEs with variable coefficients in 2D and 3D bounded and unbounded domains has been studied. In addition, direct segregated BDIE systems for Stokes system PDEs with variable coefficients in two- and three-dimensional bounded domains are presented but not for 2D unbounded cases. For a more detailed description, we refer the reader to [1, 2, 5, 6, 8, 9].

The aim of this work is to give an overview of how to derive and investigate the equivalence of BDIE systems to the mixed BVPs for compressible Stokes equations with variable viscosity in a 2D unbounded domain.

2 Preliminaries

Let $\Omega := \Omega^+$ be a unbounded (exterior) simply connected domain in $\mathbb{R}^2$ and let $\Omega^- := \mathbb{R}^2 \setminus \overline{\Omega^+}$ be the complementary (bounded) subset of Ω. The boundary $\partial\Omega$ is simply connected, closed, and $\mathcal{C}^2$–smooth for simplicity.

In what follows, we use the bold notation: $\boldsymbol{H}^s(\Omega) = [H^s(\Omega)]^2$ for two-dimensional vector spaces. We denote by $\widetilde{\boldsymbol{H}}^s(\Omega)$ the subspace of $\boldsymbol{H}^s(\mathbb{R}^2)$ defined as $\widetilde{\boldsymbol{H}}^s(\Omega) := \{\boldsymbol{g} : \boldsymbol{g} \in \boldsymbol{H}^s(\mathbb{R}^2), \operatorname{supp} \boldsymbol{g} \subset \overline{\Omega}\}$; similarly, $\widetilde{\boldsymbol{H}}^s(S_1) = \{\boldsymbol{g} \in \boldsymbol{H}^s(\partial\Omega), \operatorname{supp} \boldsymbol{g} \subset \overline{S_1}\}$ is the Sobolev space of functions having support in $S_1 \subset \partial\Omega$. We will use the following notation for derivative operators: $\partial_j = \partial_{x_j} := \frac{\partial}{\partial x_j}$ with $j = 1, 2$; $\nabla := (\partial_1, \partial_2)$.

Furthermore, to ensure the unique solvability of the BVPs in exterior domains, we will need the *weighted Sobolev spaces*; see, e.g. [3, 7]. Let us first introduce the weighted Lebesgue space:

$$L_2(\rho^{-1}; \Omega) = \{g : \rho^{-1} g \in L_2(\Omega)\}, \quad \text{where } \rho(\boldsymbol{x}) = (1 + |\mathbf{x}|^2)^{1/2} \ln(2 + |\mathbf{x}|^2).$$

Let $\mathcal{H}^1(\Omega)$ denote the following weighted Sobolev (Beppo-Levi) space: constructed using the $L_2(\rho^{-1}; \Omega)$ space

$$\mathcal{H}^1(\Omega) := \{g \in L_2(\rho^{-1}; \Omega) : \nabla g \in L_2(\Omega)\}$$

endowed with the corresponding norm $\|g\|^2_{\mathcal{H}^1(\Omega)} := \|\rho^{-1}g\|^2_{L_2(\Omega)} + \|\nabla g\|^2_{L_2(\Omega)}$. The analogous vector counterpart of $\mathcal{H}^1(\Omega)$ reads $\boldsymbol{\mathcal{H}}^1(\Omega) := \{\boldsymbol{g} \in \boldsymbol{L}_2(\rho^{-1};\Omega) : \operatorname{grad}\boldsymbol{g} \in L_2(\Omega)^{2\times 2}\}$.

If Ω^- is bounded, then $\boldsymbol{\mathcal{H}}^1(\Omega^-) = \boldsymbol{H}^1(\Omega^-)$. If Ω' is a bounded subdomain of an unbounded domain Ω and $\boldsymbol{g} \in \boldsymbol{\mathcal{H}}^1(\Omega)$, then $\boldsymbol{g} \in \boldsymbol{H}^1(\Omega')$.

Let $\widetilde{\boldsymbol{\mathcal{H}}}^1(\Omega)$ be the completion of $\boldsymbol{\mathcal{D}}(\Omega)$ in $\boldsymbol{\mathcal{H}}^1(\mathbb{R}^2)$; it can be also characterised as $\widetilde{\boldsymbol{\mathcal{H}}}^1(\Omega) = \{\boldsymbol{g} : \boldsymbol{g} \in \boldsymbol{\mathcal{H}}^1(\mathbb{R}^2),\ \operatorname{supp}\boldsymbol{g} \subset \overline{\Omega}\}$. Let $\widetilde{\boldsymbol{\mathcal{H}}}^{-1}(\Omega) := [\boldsymbol{\mathcal{H}}^1(\Omega)]^*$ and $\boldsymbol{\mathcal{H}}^{-1}(\Omega) := [\widetilde{\boldsymbol{\mathcal{H}}}^1(\Omega)]^*$ be the corresponding dual spaces. Evidently, the space $\boldsymbol{L}_2(\rho;\Omega) \subset \boldsymbol{\mathcal{H}}^{-1}(\Omega)$ and $\mathbf{H}_{**}^{-\frac{1}{2}}(\partial\Omega) := \{\boldsymbol{\rho} \in \mathbf{H}^{-\frac{1}{2}}(\partial\Omega) : \langle\rho_i, 1\rangle_{\partial\Omega} = 0 \quad \text{for} \quad i = 1, 2\}$.

Let μ be the viscosity coefficient, p the pressure field, and $\boldsymbol{v}$ the velocity field. In this paper, for an arbitrary couple $(p, \boldsymbol{v})$, the stress tensor operator, σ_{ij}, and the Stokes operator, $\mathcal{A}_j$, are defined for a compressible fluid as

$$\sigma_{ji}(p,\boldsymbol{v})(\boldsymbol{x}) := -\,\delta_i^j p(\boldsymbol{x}) + \mu(\boldsymbol{x})\left(\frac{\partial v_i(\boldsymbol{x})}{\partial x_j} + \frac{\partial v_j(\boldsymbol{x})}{\partial x_i} - \alpha\delta_i^j \operatorname{div}\boldsymbol{v}(\boldsymbol{x})\right), \tag{2.1}$$

$$\mathcal{A}_j(p,\boldsymbol{v})(\boldsymbol{x}) := \frac{\partial}{\partial x_i}\sigma_{ji}(p,\boldsymbol{v})(\boldsymbol{x}) = \frac{\partial}{\partial x_i}\left(\mu(\boldsymbol{x})\left(\frac{\partial v_j(\boldsymbol{x})}{\partial x_i} + \frac{\partial v_i(\boldsymbol{x})}{\partial x_j} - \alpha\delta_i^j \operatorname{div}\boldsymbol{v}(\boldsymbol{x})\right)\right) - \frac{\partial p(\boldsymbol{x})}{\partial x_j},\ j, i \in \{1, 2\}, \tag{2.2}$$

where $\alpha = 1$ or $\alpha = \frac{2}{3}$ and δ_i^j is Kronecker symbol. Henceforth we assume the Einstein summation in repeated indices from 1 to 2 if not stated otherwise.

Condition 2.1

$$\mu \in C^1(\mathbb{R}^2) \cap L_\infty(\mathbb{R}^2) \ :\ \rho\nabla\mu \in \boldsymbol{L}_\infty(\mathbb{R}^2).$$

In addition, there exist constants C_1 and C_2 such that

$$0 < C_1 < \mu(\boldsymbol{x}) < C_2.$$

Remark 2.2 If μ satisfies condition 2.1, then the functions μ and $\dfrac{1}{\mu}$ are multipliers in the space $\mathcal{H}^1(\Omega)$, i.e. there exists positive constants $C_3(\mu)$ and $C_4(\mu)$ independent of h such that

$$\|\mu h\|_{\mathcal{H}^1(\Omega)} \le C_3(\mu)\|h\|_{\mathcal{H}^1(\Omega)}, \quad \left\|\frac{h}{\mu}\right\|_{\mathcal{H}^1(\Omega)} \le C_4(\mu)\|h\|_{\mathcal{H}^1(\Omega)}, \quad \forall h \in \mathcal{H}^1(\Omega). \tag{2.3}$$

The operator $\mathcal{A}$ acting on $(p, \boldsymbol{v}) \in L_2(\Omega) \times \mathcal{H}^1(\Omega)$ is well defined in the weak sense as long as the variable coefficient $\mu(\boldsymbol{x})$ is essentially bounded, i.e. $\mu \in L_\infty(\Omega)$. Indeed, in the sense of distributions, the operator $\mathcal{A}$ is defined as

$$\langle \mathcal{A}(p, \boldsymbol{v}), \boldsymbol{u} \rangle_\Omega = -\mathcal{E}((p, \boldsymbol{v}), \boldsymbol{u}), \quad \forall \boldsymbol{u} \in \mathcal{D}(\Omega), \tag{2.4}$$

where

$$\mathcal{E}\left((p, \boldsymbol{v}), \boldsymbol{u}\right) := \int_\Omega E\left((p, \boldsymbol{v}), \boldsymbol{u}\right)(\boldsymbol{x})\, dx, \tag{2.5}$$

and the function $E\left((p, \boldsymbol{v}), \boldsymbol{u}\right)$ is defined as

$$E\left((p, \boldsymbol{v}), \boldsymbol{u}\right)(\boldsymbol{x}) := \frac{1}{2}\mu(\boldsymbol{x}) \left(\frac{\partial u_i(\boldsymbol{x})}{\partial x_j} + \frac{\partial u_j(\boldsymbol{x})}{\partial x_i} \right) \left(\frac{\partial v_i(\boldsymbol{x})}{\partial x_j} + \frac{\partial v_j(\boldsymbol{x})}{\partial x_i} \right)$$
$$- \alpha \mu(\boldsymbol{x}) \mathrm{div} \boldsymbol{v}(\boldsymbol{x})\, \mathrm{div} \boldsymbol{u}(\boldsymbol{x}) - p(\boldsymbol{x}) \mathrm{div} \boldsymbol{u}(\boldsymbol{x}).$$

We will also make use of the following space (cf., e.g. [7]):

$$\mathcal{H}^{1,0}(\Omega; \mathcal{A}) := \{(p, \boldsymbol{v}) \in L_2(\Omega) \times \mathcal{H}^1(\Omega) : \mathcal{A}(p, \boldsymbol{v}) \in \boldsymbol{L}_2(\rho; \Omega)\},$$

endowed with the norm, $\|\cdot\|_{\mathcal{H}^{1,0}(\Omega; \mathcal{A})}$, where $\|(p, \boldsymbol{v})\|_{\mathcal{H}^{1,0}(\Omega; \mathcal{A})} := \Big(\|p\|^2_{L_2(\Omega)} + \|\boldsymbol{v}\|^2_{\mathcal{H}^1(\Omega)} + \|\rho \mathcal{A}(p, \boldsymbol{v})\|^2_{\boldsymbol{L}_2(\Omega)} \Big)^{1/2}$.

Theorem 2.3 *Let μ satisfy the condition 2.1. Then the space $\mathcal{D}(\overline{\Omega}) \times \mathcal{D}(\overline{\Omega})$ is dense in $\mathcal{H}^{1,0}(\Omega; \mathcal{A})$.*

For sufficiently smooth functions $(p, \boldsymbol{v}) \in H^{s-1}(\Omega^\pm) \times \boldsymbol{H}^s(\Omega^\pm)$ with $s > 3/2$, we can define the classical traction (conormal derivative) operators $\boldsymbol{T}^{c\pm} = \{T_i^{c\pm}\}_{i=1}^2$, on the boundary $\partial\Omega$ as

$$T_i^{c\pm}(p, \boldsymbol{v})(\boldsymbol{x}) := [\gamma^\pm \sigma_{ij}(p, \boldsymbol{v})(\boldsymbol{x})] n_j(\boldsymbol{x}), \tag{2.6}$$

where $n_j(\boldsymbol{x})$ denotes the components of the unit normal vector $\boldsymbol{n}(\boldsymbol{x})$ to the boundary $\partial\Omega$ *directed outwards the exterior domain* Ω. Moreover, $\boldsymbol{\gamma}^\pm$ denotes the trace operators from inside and outside Ω.

The traction operators (2.6) can be continuously extended to the *canonical* traction operators $\boldsymbol{T}^\pm : \mathcal{H}^{1,0}(\Omega^\pm, \mathcal{A}) \to \boldsymbol{H}^{-1/2}(\partial\Omega)$ defined in the weak form (cf. [7, 9]), as

$$\langle \boldsymbol{T}^+(p, \boldsymbol{v}), \boldsymbol{w} \rangle_{\partial\Omega} := \int_{\Omega^\pm} \left[\mathcal{A}(p, \boldsymbol{v}) \boldsymbol{\gamma}_{-1}^+ \boldsymbol{w} + E\left((p, \boldsymbol{v}), \boldsymbol{\gamma}_{-1}^+ \boldsymbol{w}\right) \right] dx$$
$$\forall (p, \boldsymbol{v}) \in \mathcal{H}^{1,0}(\Omega^\pm, \mathcal{A}),\ \forall \boldsymbol{w} \in \boldsymbol{H}^{1/2}(\partial\Omega),$$

where the operator $\gamma^{+}_{-1} : H^{1/2}(\partial\Omega) \to \mathcal{H}^{1}(\Omega)$ denotes a continuous right inverse of the trace operator $\gamma^{+} : \mathcal{H}^{1}(\Omega) \to H^{1/2}(\partial\Omega)$.

Furthermore, if $(p, v) \in \mathcal{H}^{1,0}(\Omega, \mathcal{A})$ and $u \in \mathcal{H}^{1}(\Omega)$, the following first Green identity holds:

$$\langle T^{+}(p, v), \gamma^{+}u\rangle_{\partial\Omega} = \int_{\Omega} [\mathcal{A}(p, v)u + E\left((p, v), u\right)(x)]dx. \tag{2.7}$$

Applying the identity (2.7) to the pairs $(p, v), (q, u) \in \mathcal{H}^{1,0}(\Omega, \mathcal{A})$ with exchanged roles and subtracting one from the other, we arrive at the second Green identity, cf. [12]:

$$\begin{aligned}&\langle T^{+}(p, v), \gamma^{+}u\rangle_{\partial\Omega} - \langle T^{+}(q, u), \gamma^{+}v\rangle_{\partial\Omega}\\ &= \int_{\Omega} \Big[\mathcal{A}_j(p, v)u_j - \mathcal{A}_j(q, u)v_j + q \operatorname{div} v - p \operatorname{div} u\Big] dx.\end{aligned} \tag{2.8}$$

In this short paper, we give an overview of how we derive systems of boundary-domain integral equations, which are equivalent to the following mixed compressible exterior Stokes problem.

For $f \in L_2(\rho, \Omega)$, $g \in L^2(\Omega)$, $\varphi_0 \in H^{1/2}(\partial\Omega_D)$, and $\psi_0 \in H^{-1/2}(\partial\Omega_N)$, find $(p, v) \in \mathcal{H}^{1,0}(\Omega, \mathcal{A})$ such that

$$\mathcal{A}(p, v)(x) = f(x), \quad x \in \Omega, \tag{2.9a}$$

$$\operatorname{div}(v)(x) = g(x), \quad x \in \Omega, \tag{2.9b}$$

$$r_{\partial\Omega_D}\gamma^{+}v(x) = \varphi_0(x), \quad x \in \partial\Omega_D, \tag{2.9c}$$

$$r_{\partial\Omega_N}T^{+}(p, v)(x) = \psi_0(x), \quad x \in \partial\Omega_N. \tag{2.9d}$$

To prove the following theorem, the procedure is similar to as in [5, 13].

Theorem 2.4 *Let μ satisfy condition 2.1;the mixed BVP (2.9) has at most one solution in space $\mathcal{H}^{1,0}(\Omega, \mathcal{A})$.*

3 Parametrix and Remainder

The operator $\mathcal{A}$ becomes the constant-coefficient of Stokes operator $\mathring{\mathcal{A}}$ when $\mu = 1$. The fundamental solution defined by the pair of distributions $(\mathring{q}^k, \mathring{\mathbf{u}}^k)$,where $\mathring{u}^k_j$ and $\mathring{q}^k$ components a fundamental solution (see, e.g.[10, 11]). So for $r_0 > 0$, $(\mathring{q}^k, \mathring{\mathbf{u}}^k)$

will have the form

$$(\mathring{q}^k(\mathbf{x},\mathbf{y}),\ \mathring{u}^k_j(\mathbf{x},\mathbf{y}))$$
$$=\left(\frac{1}{4\pi}\left(\frac{-(x_k-y_k)}{2\pi|\mathbf{x}-\mathbf{y}|^2},\ \delta^k_j \log\frac{|\mathbf{x}-\mathbf{y}|}{r_0}-\frac{(x_j-y_j)(x_k-y_k)}{|\mathbf{x}-\mathbf{y}|^2}\right)\right).$$

Let us denote $\mathring{\sigma}_{ij}(p,\mathbf{v}) := \sigma_{ij}(p,\mathbf{v})|_{\mu=1}$. Then in a particular case, for $\mu = 1$ and the fundamental solution $(\mathring{q}^k, \mathring{\mathbf{u}}^k)_{k=1,2}$ of the operator $\mathring{\mathcal{A}}$, the stress tensor $\mathring{\sigma}_{ij}(\mathring{q}^k, \mathring{\mathbf{u}}^k)(\mathbf{x}-\mathbf{y})$ reads as

$$\mathring{\sigma}_{ij}(\mathbf{x};\mathring{q}^k,\mathring{\mathbf{u}}^k)(\mathbf{x}-\mathbf{y}) = \frac{1}{\pi}\frac{(x_i-y_i)(x_j-y_j)(x_k-y_k)}{|\mathbf{x}-\mathbf{y}|^4}.$$

The boundary traction becomes

$$\mathring{T}^c_j(\mathbf{x};\mathring{q}^k,\mathring{\mathbf{u}}^k)(\mathbf{x},\mathbf{y}) := \mathring{\sigma}_{ij}(\mathring{q}^k,\mathring{u}^k)(\mathbf{x}-\mathbf{y})n_i(\mathbf{x})$$
$$= \frac{1}{\pi}\frac{(x_i-y_i)(x_j-y_j)(x_k-y_k)}{|\mathbf{x}-\mathbf{y}|^4}n_i(\mathbf{x}).$$

Let us define a pair of functions $(q^k, \mathbf{u}^k)_{k=1,2}$ similar as in [5]:

$$u^k_j(\mathbf{x},\mathbf{y}) = \frac{1}{\mu(\mathbf{y})}\mathring{u}_j^k(\mathbf{x},\mathbf{y}) = \frac{1}{4\pi\mu(\mathbf{y})}\left(\delta^k_j\log\frac{|\mathbf{x}-\mathbf{y}|}{r_0}-\frac{(x_j-y_j)(x_k-y_k)}{|\mathbf{x}-\mathbf{y}|^2}\right), \tag{3.1}$$

$$q^k(\mathbf{x},\mathbf{y}) = \frac{\mu(\mathbf{x})}{\mu(\mathbf{y})}\mathring{q}^k(\mathbf{x},\mathbf{y}) = \frac{\mu(\mathbf{x})}{\mu(\mathbf{y})}\frac{y_k-x_k}{2\pi|\mathbf{x}-\mathbf{y}|^2},\quad j,k\in\{1,2\}. \tag{3.2}$$

Then substituting (3.1)–(3.2) into the Stokes system (2.9a) with variable coefficients, we get

$$\mathcal{A}_j(\mathbf{x};q^k;\mathbf{u}^k)(\mathbf{x},\mathbf{y}) = \delta^k_j\delta(\mathbf{x}-\mathbf{y}) + R_{kj}(\mathbf{x},\mathbf{y}), \tag{3.3}$$

where

$$R_{kj}(\mathbf{x},\mathbf{y}) = \frac{1}{\mu(\mathbf{y})}\frac{\partial(\mu(\mathbf{x}))}{\partial x_i}\mathring{\sigma}_{ij}(\mathring{q}^k,\mathring{\mathbf{u}}^k)(\mathbf{x}-\mathbf{y}) = \mathcal{O}(|\mathbf{x}-\mathbf{y}|^{-1})$$

is a weakly singular remainder. This implies that $(q^k,\ \mathbf{u}^k)$ is a parametrix of the operator $\mathcal{A}$.

3.1 Volume and Surface Potentials

Let ρ and $\boldsymbol{\rho}$ be sufficiently smooth scalar and vector functions on Ω. The parametrix-based Newton-type and the Remainder vector potential operators are defined as

$$[\boldsymbol{\mathcal{U}}\boldsymbol{\rho}]_k(\mathbf{y}) = \mathcal{U}_{kj}\rho_j(\mathbf{y}) := \int_\Omega u_j^k(\mathbf{x},\mathbf{y})\rho_j(\mathbf{x})d\mathbf{x},$$

$$[\boldsymbol{\mathcal{R}}\boldsymbol{\rho}]_k(\mathbf{y}) = \mathcal{R}_{kj}\rho_j(\mathbf{y}) := \int_\Omega R_{kj}(\mathbf{x},\mathbf{y})\rho_j(\mathbf{x})d\mathbf{x}, \quad \mathbf{y} \in \mathbb{R}^2$$

for the velocity $\mathbf{v}$, and the Newton-type scalar and remainder potentials for the pressure,

$$[\boldsymbol{\mathcal{Q}}\rho]_j(\mathbf{y}) = \mathcal{Q}_j\rho(\mathbf{y}) := -\int_\Omega q^j(\mathbf{x},\mathbf{y})\rho(\mathbf{x})d\mathbf{x},$$

$$\mathcal{Q}\boldsymbol{\rho}(\mathbf{y}) = \boldsymbol{\mathcal{Q}}\cdot\boldsymbol{\rho}(\mathbf{y}) = \mathcal{Q}_j\rho_j(\mathbf{y}) := -\int_\Omega q^j(\mathbf{x},\mathbf{y})\rho_j(\mathbf{x})d\mathbf{x},$$

$$\begin{aligned}\boldsymbol{\mathcal{R}}^\bullet\boldsymbol{\rho}(\mathbf{y}) &= -2\langle\partial_i\mathring{q}^j(\cdot,\mathbf{y}),\rho_i\partial_j\mu\rangle_\Omega - 2\rho_i(\mathbf{y})\partial_i\mu(\mathbf{y})\\ &= -2v.p.\int_\Omega \frac{\partial\mathring{q}^j(\mathbf{x},\mathbf{y})}{\partial x_i}\frac{\partial\mu(\mathbf{x})}{\partial x_i}\rho_j(\mathbf{x})d\mathbf{x} - \rho_j(\mathbf{y})\frac{\partial\mu(\mathbf{y})}{\partial y_j},\end{aligned} \tag{3.4}$$

for $\mathbf{y} \in \mathbf{R}^2$. The integral in (3.4) is understood as a strongly singular 2D integral in the Cauchy sense (see, e.g. [2]).

For the velocity, the parametrix-based single- and double-layer potentials are defined for $\mathbf{y} \notin \partial\Omega$ as

$$[\mathbf{V}\boldsymbol{\rho}]_k(\mathbf{y}) = V_{kj}\rho_j(\mathbf{y}) := -\int_{\partial\Omega} u_j^k(\mathbf{x},\mathbf{y})\rho_j(\mathbf{x})dS_\mathbf{x},$$

$$[\mathbf{W}\boldsymbol{\rho}]_k(\mathbf{y}) = W_{kj}\rho_j(\mathbf{y}) := -\int_{\partial\Omega} T_j^+(\mathbf{x}; q^k, \mathbf{u}^k)(\mathbf{x},\mathbf{y})\rho_j(\mathbf{x})dS_\mathbf{x},$$

and for pressure in the variable coefficient Stokes system, the single-layer and double-layer potentials are defined for $\mathbf{y} \notin \partial\Omega$ as

$$\Pi^s\boldsymbol{\rho}(\mathbf{y}) = \Pi_j^s\rho_j(\mathbf{y}) := \int_{\partial\Omega} \mathring{q}^j(\mathbf{x},\mathbf{y})\rho_j(\mathbf{x})dS_\mathbf{x},$$

$$\Pi^d\boldsymbol{\rho}(\mathbf{y}) = \Pi_j^d\rho_j(\mathbf{y}) := 2\int_{\partial\Omega} \frac{\partial\mathring{q}^j(\mathbf{x},\mathbf{y})}{\partial\mathbf{n}(\mathbf{x})}\mu(\mathbf{x})\rho_j(\mathbf{x})dS_\mathbf{x}.$$

The corresponding boundary integral (pseudo-differential) operators of direct surface values of the single-layer potential and the double-layer potential and the traction of the single-layer potential and the double-layer potential are

$$[\mathcal{V}\rho]_k(\mathbf{y}) = \mathcal{V}_{kj}\rho_j(\mathbf{y}) := -\int_{\partial\Omega} u_j^k(\mathbf{x},\mathbf{y})\rho_j(\mathbf{x})dS_{\mathbf{x}},$$

$$[\mathcal{W}\rho]_k(\mathbf{y}) = \mathcal{W}_{kj}\rho_j(\mathbf{y}) := -\int_{\partial\Omega} T_j^+(\mathbf{x}; q^k, \mathbf{u}^k)(\mathbf{x},\mathbf{y})\rho_j(\mathbf{x})dS_{\mathbf{x}},$$

$$\left[\mathcal{W}'\rho\right]_k(\mathbf{y}) = \mathcal{W}'_{kj}\rho_j(\mathbf{y}) := -\int_{\partial\Omega} T_j^+(\mathbf{y}; q^k, \mathbf{u}^k)(\mathbf{x},\mathbf{y})\rho_j(\mathbf{x})dS_{\mathbf{x}},$$

$$\mathcal{L}^{\pm}\rho(\mathbf{y}) := \mathbf{T}^{\pm}(\Pi^d\rho, \mathbf{W}\rho)(\mathbf{y}), \quad \mathbf{y} \in \partial\Omega,$$

where $\mathbf{T}^{\pm}$ are the traction operators (see, e.g. [2, 4]).

Remark 3.1 The detailed analysis of the mapping properties of potentials is carried out in [1, 13].

Lemma 3.2 *Let* $\partial\Omega = \overline{S_1} \cup \overline{S_2}$, *where* S_1 *ad* S_2 *are open nonempty nonintersecting parts of the boundary curve* $\partial\Omega$. *Let* $\boldsymbol{\Psi}^* \in \widetilde{\boldsymbol{H}}_{**}^{-\frac{1}{2}}(S_1)$, $\boldsymbol{\Phi}^* \in \widetilde{\boldsymbol{H}}^{\frac{1}{2}}(S_2)$. *If*

$$\boldsymbol{V\Psi}^* - \boldsymbol{W\Phi}^* = \boldsymbol{0}, \qquad \Pi^s\boldsymbol{\Psi}^* - \Pi^d\boldsymbol{\Phi}^* = 0, \quad in\ \ \Omega,$$

then $\boldsymbol{\Psi}^* = \boldsymbol{0}$ *and* $\boldsymbol{\Phi}^* = \boldsymbol{0}$ *on* $\partial\Omega$.

4 The Third Green Identities

Theorem 4.1 *For any* $(p, \boldsymbol{v}) \in \mathcal{H}^{1,0}(\Omega; \boldsymbol{\mathcal{A}})$, *the following third Green identities hold [5, 13]:*

$$\boldsymbol{v} + \mathcal{R}\boldsymbol{v} - \boldsymbol{V}\boldsymbol{T}^+(p,\boldsymbol{v}) + \boldsymbol{W}\gamma^+\boldsymbol{v} = \boldsymbol{\mathcal{U}\mathcal{A}}(p,\boldsymbol{v}) - \boldsymbol{\mathcal{Q}}\mathrm{div}\boldsymbol{v} \ \text{ in } \Omega, \tag{4.1}$$

$$p + \mathcal{R}^{\bullet}\boldsymbol{v} - \Pi^s\boldsymbol{T}^+(p,\boldsymbol{v}) + \Pi^d\gamma^+\boldsymbol{v} = \mathring{\mathcal{Q}}\boldsymbol{\mathcal{A}}(p,\boldsymbol{v}) + (2-\alpha)\mu\mathrm{div}\boldsymbol{v} \ \text{ in } \Omega. \tag{4.2}$$

If the couple $(p, \mathbf{v}) \in \boldsymbol{\mathcal{H}}^{1,0}(\Omega; \boldsymbol{\mathcal{A}})$ is a solution of the Stokes PDE (2.9a)–(2.9b) with variable coefficient, then (4.1) and (4.2) give

$$\mathbf{v} + \boldsymbol{\mathcal{R}}\mathbf{v} - \mathbf{V}\mathbf{T}^+(p,\mathbf{v}) + \mathbf{W}\gamma^+\mathbf{v} = \boldsymbol{\mathcal{U}}\mathbf{f} - \boldsymbol{\mathcal{Q}}g, \ \text{ in } \Omega, \tag{4.3}$$

$$p + \mathcal{R}^{\bullet}\mathbf{v} - \Pi^s\mathbf{T}^+(p,\mathbf{v}) + \Pi^d\gamma^+\mathbf{v} = \mathring{\mathcal{Q}}\mathbf{f} + (2-\alpha)\mu g, \ \text{ in } \Omega. \tag{4.4}$$

We will also need the trace and traction of the third Green identities (4.3) and (4.4) on $\partial\Omega$.

$$\frac{1}{2}\gamma^+\mathbf{v} + \mathcal{R}^+\mathbf{v} - \mathcal{V}\mathbf{T}^+(p, \mathbf{v}) + \mathcal{W}\gamma^+\mathbf{v} = \gamma^+\mathcal{U}\mathbf{f} - \gamma^+\mathcal{Q}g, \tag{4.5}$$

$$\frac{1}{2}\mathbf{T}^+(p, \mathbf{v}) + \mathbf{T}^+(\mathcal{R}^\bullet, \mathcal{R})\mathbf{v} - \mathcal{W}'\mathbf{T}^+(p, \mathbf{v}) + \mathcal{L}^+\gamma^+\mathbf{v}$$
$$= \mathbf{T}^+(\mathring{\mathcal{Q}}\mathbf{f} + (2-\alpha)\mu g, \mathcal{U}\mathbf{f} - \mathcal{Q}g). \tag{4.6}$$

Condition 4.2

$$\mu \in \mathcal{C}^2(\mathbb{R}^2) \ : \ \rho^2\partial_j\partial_i\mu \in L_\infty(\mathbb{R}^2). \tag{4.7}$$

The following theorem is instrumental for proof of equivalence of the BDIEs and the original PDE.

Lemma 4.3 *Let conditions 2.1 and 4.2 hold. Let* $\mathbf{v} \in \mathcal{H}^1(\Omega)$, $p \in L^2(\Omega)$, $g \in L^2(\Omega)$, $\boldsymbol{f} \in \boldsymbol{L}^2(\rho; \Omega)$, $\boldsymbol{\Psi} \in \boldsymbol{H}^{-\frac{1}{2}}(\partial\Omega)$, $\boldsymbol{\Phi} \in \boldsymbol{H}^{\frac{1}{2}}(\partial\Omega)$ *satisfy equations.*

$$\boldsymbol{v} + \mathcal{R}\boldsymbol{v} - \boldsymbol{V}\boldsymbol{\Psi} + \boldsymbol{W}\boldsymbol{\Phi} = \mathcal{U}\boldsymbol{f} - \mathcal{Q}g,$$
$$p + \mathcal{R}^\bullet\boldsymbol{v} - \Pi^s\boldsymbol{\Psi} + \Pi^d\boldsymbol{\Phi} = \mathring{\mathcal{Q}}\boldsymbol{f} + (2-\alpha)\mu g, \ \text{ in } \Omega.$$

Then $(p, \boldsymbol{v}) \in \mathcal{H}^{1,0}(\Omega; \boldsymbol{\mathcal{A}})$ *and solve the equations*

$$\boldsymbol{\mathcal{A}}(\boldsymbol{y}; p, \boldsymbol{v}) = \boldsymbol{f}, \ \operatorname{div}\boldsymbol{v} = g.$$

Moreover, the following relations hold true:

$$\boldsymbol{V}(\boldsymbol{\Psi} - \boldsymbol{T}^+(p, \boldsymbol{v}))(\boldsymbol{y}) - \boldsymbol{W}(\boldsymbol{\Phi} - \gamma^+\boldsymbol{v})(\boldsymbol{y}) = 0,$$
$$\Pi^s(\boldsymbol{\Psi} - \boldsymbol{T}^+(p, \boldsymbol{v}))(\boldsymbol{y}) - \Pi^d(\boldsymbol{\Phi} - \gamma^+\boldsymbol{v})(\boldsymbol{y}) = 0, \quad \boldsymbol{y} \in \Omega.$$

5 BDIEs for Mixed BVP

We aim to obtain four different segregated boundary-domain integral equation systems for mixed BVP (2.9a)–(2.9d). We will use similar procedures as in [5, 13]. To this end, let $\boldsymbol{\Phi}_0 \in \mathbf{H}^{\frac{1}{2}}(\partial\Omega)$ be a fixed extension of the given data $\boldsymbol{\varphi}_0$ from $\partial\Omega_D$ to $\partial\Omega$. An arbitrary extension $\boldsymbol{\Phi} \in \mathbf{H}^{\frac{1}{2}}(\partial\Omega)$ preserving the function space can be represented as $\boldsymbol{\Phi} = \boldsymbol{\Phi}_0 + \boldsymbol{\varphi}$ with $\boldsymbol{\varphi} \in \widetilde{\mathbf{H}}^{\frac{1}{2}}(\partial\Omega_N)$. Analogously, let $\boldsymbol{\Psi}_0 \in \mathbf{H}^{-\frac{1}{2}}(\partial\Omega)$ be a fixed extension of the given data $\boldsymbol{\psi}_0$ from $\partial\Omega_N$ to $\partial\Omega$.

An arbitrary extension $\boldsymbol{\Psi} \in \mathbf{H}^{-\frac{1}{2}}(\partial\Omega)$ preserving the function space can then be represented as $\boldsymbol{\Psi} = \boldsymbol{\Psi}_0 + \boldsymbol{\psi}$ with $\boldsymbol{\psi} \in \widetilde{\mathbf{H}}^{-\frac{1}{2}}(\partial\Omega_D)$. Let us now represent

$$\gamma^+\mathbf{v} = \boldsymbol{\Phi}_0 + \boldsymbol{\varphi}, \quad \mathbf{T}^+(p, \mathbf{v}) = \boldsymbol{\Psi}_0 + \boldsymbol{\psi} \ \ on \ \ \partial\Omega, \tag{5.1}$$

where $\boldsymbol{\varphi} \in \widetilde{\mathbf{H}}^{\frac{1}{2}}(\partial\Omega_N)$ and $\boldsymbol{\psi} \in \widetilde{\mathbf{H}}^{-\frac{1}{2}}(\partial\Omega_D)$ are unknown boundary functions.

BDIE system (M11) Let us now take Eqs. (4.3) and (4.4) in Ω and restrictions of Eqs. (4.5) and (4.6) to the boundary parts $\partial\Omega_D$ and $\partial\Omega_N$, respectively. We obtain the following BDIE system **(M11)** consisting of four BDIEs for four unknowns, $(p, \mathbf{v}, \boldsymbol{\psi}, \boldsymbol{\varphi}) \in \mathcal{H}^{1,0}(\Omega; \boldsymbol{\mathcal{A}}) \times \widetilde{\mathbf{H}}^{-\frac{1}{2}}(\partial\Omega_D) \times \widetilde{\mathbf{H}}^{\frac{1}{2}}(\partial\Omega_N)$:

$$p + \mathcal{R}^\bullet\mathbf{v} - \Pi^s\boldsymbol{\psi} + \Pi^d\boldsymbol{\varphi} = F_0 \ \ \text{in} \ \ \Omega, \tag{5.2a}$$

$$\mathbf{v} + \boldsymbol{\mathcal{R}}\mathbf{v} - \mathbf{V}\boldsymbol{\psi} + \mathbf{W}\boldsymbol{\varphi} = \mathbf{F} \ \ \text{in} \ \ \Omega, \tag{5.2b}$$

$$r_{\partial\Omega_D}\gamma^+\mathcal{R}\mathbf{v} - r_{\partial\Omega_D}\boldsymbol{\mathcal{V}}\boldsymbol{\psi} + r_{\partial\Omega_D}\boldsymbol{\mathcal{W}}\boldsymbol{\varphi} = r_{\partial\Omega_D}\gamma^+\mathbf{F} - \boldsymbol{\varphi}_0 \ \ \text{on} \ \ \partial\Omega_D, \tag{5.2c}$$

$$r_{\partial\Omega_N}\mathbf{T}^+(\mathcal{R}^\bullet, \boldsymbol{\mathcal{R}})\mathbf{v} - r_{\partial\Omega_N}\boldsymbol{\mathcal{W}}'\boldsymbol{\psi} + r_{\partial\Omega_N}\boldsymbol{\mathcal{L}}^+\boldsymbol{\varphi} = r_{\partial\Omega_N}\mathbf{T}^+(F_0, \mathbf{F}) - \boldsymbol{\psi}_0 \ \ \text{on} \ \ \partial\Omega_N, \tag{5.2d}$$

where

$$F_0 := \mathring{\mathcal{Q}}\mathbf{f} - (2-\alpha)\mu g + \Pi^s\boldsymbol{\Psi}_0 - \Pi^d\boldsymbol{\Phi}_0, \quad \mathbf{F} := \boldsymbol{\mathcal{U}}\mathbf{f} - \boldsymbol{\mathcal{Q}}g + \mathbf{V}\boldsymbol{\Psi}_0 - \mathbf{W}\boldsymbol{\Phi}_0. \tag{5.3}$$

We denote the right-hand side of BDIE system (5.2a)–(5.2d) as $\mathcal{F}^{11} := [F_0, \mathbf{F}, \ r_{\partial\Omega_D}\gamma^+\mathbf{F} - \boldsymbol{\varphi}_0, \ r_{\partial\Omega_N}T^+(F_0, \mathbf{F}) - \boldsymbol{\psi}_0]^T$, which implies $\mathcal{F}^{11} \in \mathcal{H}^{1,0}(\Omega; \boldsymbol{\mathcal{A}}) \times \mathbf{H}^{\frac{1}{2}}(\partial\Omega_D) \times \mathbf{H}^{-\frac{1}{2}}(\partial\Omega_N)$. The term $\mathcal{F}^{11} = \mathbf{0}$ if and only if $(\mathbf{f}, g, \boldsymbol{\Phi}_0, \boldsymbol{\Psi}_0) = \mathbf{0}$.

BDIE system (M12) Let us now take Eqs. (4.3) and (4.4) in Ω and Eq. (4.5) on the whole boundary $\partial\Omega$. Substituting there representations (5.1), we arrive at the following system of BDIEs:

$$p + \mathcal{R}^\bullet\mathbf{v} - \Pi^s\boldsymbol{\psi} + \Pi^d\boldsymbol{\varphi} = F_0 \ \ \text{in} \ \ \Omega,$$

$$\mathbf{v} + \boldsymbol{\mathcal{R}}\mathbf{v} - \mathbf{V}\boldsymbol{\psi} + \mathbf{W}\boldsymbol{\varphi} = \mathbf{F} \ \ \text{in} \ \ \Omega,$$

$$\frac{1}{2}\boldsymbol{\varphi} + \gamma^+\mathcal{R}\mathbf{v} - \boldsymbol{\mathcal{V}}\boldsymbol{\psi} + \boldsymbol{\mathcal{W}}\boldsymbol{\varphi} = \gamma^+\mathbf{F} - \boldsymbol{\Phi}_0 \ \ \text{on} \ \ \partial\Omega,$$

where F_0 and $\mathbf{F}$ are given by (5.3).

BDIE system (M21) Let us now take Eqs. (4.3) and (4.4) in Ω and Eq. (4.6) on the whole boundary $\partial\Omega$. Substituting their representations (5.1), we arrive at the following system of BDIEs:

$$p + \mathcal{R}^{\bullet}\mathbf{v} - \Pi^s\psi + \Pi^d\varphi = F_0 \quad \text{in } \Omega,$$

$$\mathbf{v} + \boldsymbol{\mathcal{R}}\mathbf{v} - \mathbf{V}\psi + \mathbf{W}\varphi = \mathbf{F} \quad \text{in } \Omega,$$

$$\frac{1}{2}\psi + \mathbf{T}^+(\mathcal{R}^{\bullet}, \boldsymbol{\mathcal{R}})\mathbf{v} - \mathcal{W}'\psi + \mathcal{L}^+\varphi = \mathbf{T}^+(F_0, \mathbf{F}) - \Psi_0 \quad \text{on } \partial\Omega,$$

where F_0 and $\mathbf{F}$ are given by (5.3).

BDIE system (M22) Let us now take Eqs. (4.3) and (4.4) in Ω and restrictions of Eqs. (4.6) and (4.5) to the boundary parts $\partial\Omega_D$ and $\partial\Omega_N$, respectively. We obtain the following BDIE system (**M22**) consisting of four BDIEs for four unknowns, $(p, \mathbf{v}) \in \mathcal{H}^{1,0}(\Omega; \mathcal{A})$, $\varphi \in \widetilde{\mathbf{H}}^{\frac{1}{2}}(\partial\Omega_N)$ and $\psi \in \widetilde{\mathbf{H}}^{-\frac{1}{2}}(\partial\Omega_D)$:

$$p + \mathcal{R}^{\bullet}\mathbf{v} - \Pi^s\psi + \Pi^d\varphi = F_0 \quad \text{in } \Omega,$$

$$\mathbf{v} + \boldsymbol{\mathcal{R}}\mathbf{v} - \mathbf{V}\psi + \mathbf{W}\varphi = \mathbf{F} \quad \text{in } \Omega,$$

$$\frac{1}{2}\psi + r_{\partial\Omega_D}[\mathbf{T}^+(\mathcal{R}^{\bullet}, \boldsymbol{\mathcal{R}})\mathbf{v} - \mathcal{W}'\psi + \mathcal{L}^+\varphi] = r_{\partial\Omega_D}[\mathbf{T}^+(F_0, \mathbf{F}) - \Psi_0] \quad \text{on } \partial\Omega_D,$$

$$\frac{1}{2}\varphi + r_{\partial\Omega_N}[\gamma^+\boldsymbol{\mathcal{R}}v - \mathcal{V}\psi + \mathcal{W}\varphi] = r_{\partial\Omega_N}[\gamma^+\mathbf{F} - \Phi_0] \quad \text{on } \partial\Omega_N,$$

where the terms on the right-hand side F_0 and $\mathbf{F}$ are given by (5.3).

Theorem 5.1 (Equivalence Theorem) *Let conditions 2.1 and 4.2 hold. Let $f \in L^2(\rho; \Omega)$, $g \in L^2(\Omega)$, and $\Phi_0 \in H^{\frac{1}{2}}(\partial\Omega)$, and $\Psi_0 \in H^{-\frac{1}{2}}(\partial\Omega)$ be some fixed extensions of $\varphi_0 \in H^{\frac{1}{2}}(\partial\Omega_D)$ and $\psi_0 \in H^{-\frac{1}{2}}(\partial\Omega_N)$ respectively.*

(i) If some $(p, v) \in L^2(\Omega) \times \mathcal{H}^1(\Omega)$ solves mixed BVP (2.9a) - (2.9d), *then the solution is unique and the set $(p, v, \psi, \varphi) \in H^{1,0}(\Omega; \mathcal{A}) \times \widetilde{H}^{-\frac{1}{2}}(\partial\Omega_D) \times \widetilde{H}^{\frac{1}{2}}(\partial\Omega_N)$, where*

$$\varphi = \gamma^+ v - \Phi_0, \quad \psi = T^+(p, v) - \Psi_0 \quad \text{on } \partial\Omega \tag{5.4}$$

*solves the BDIE system (**M11**).*

*(ii) If $(p, v, \psi, \varphi) \in L^2(\Omega) \times \mathcal{H}^1(\Omega) \times \widetilde{H}_{**}^{-\frac{1}{2}}(\partial\Omega_D) \times \widetilde{H}^{\frac{1}{2}}(\partial\Omega_N)$ solves one of the BDIE system (**M11**), then the solution is unique and solves the BDIE systems, while (p, v) belongs to $H^{1,0}(\Omega; \mathcal{A})$ and solves the mixed BVP* (2.9a)–(2.9d) *and relations* (5.4).

Acknowledgments The research is financially supported by the Ghent Analysis & PDE Center, Department of Mathematics: Analysis, Logic and Discrete Mathematics, Ghent University.

References

1. M.A. Dagnaw, C. Fresneda-Portillo, Mapping properties of potential operators related to the 2D compressible stokes system in weighted Sobolev spaces, in *Integral Methods in Science and Engineering: Applications in Theoretical and Practical Research* (Birkhäuser, Cham, 2022), pp. 67–83
2. M.A. Dagnaw, C. Fresneda-Portillo, On the well-posedness of two boundary-domain integral equation systems equivalent to the Dirichlet problem for the stokes system with variable viscosity. Results Math. **77**, 99 (2022)
3. F. Alliot, C. Amrouche, Weak solutions for the exterior Stokes problem in weighted Sobolev spaces. Math. Methods Appl. Sci. **23**, 575–600 (2000)
4. T.G. Ayele, M.A. Dagnaw, Boundary-domain integral equation systems to the Dirichlet and Neumann problems for compressible Stokes equations with variable viscosity in 2D. Math. Methods Appl. Sci. **44**, 9876–9898 (2021)
5. T.G. Ayele, M.A. Dagnaw, Boundary-domain integral equation systems to the mixed BVP for compressible Stokes equations with variable viscosity in 2D. Math. Methods Appl. Sci. **44**, 9899–9926 (2021)
6. O. Chkadua, S.E. Mikhailov, D. Natroshvili, Analysis of direct boundary-domain integral equations for a mixed BVP with variable coefficient, I: equivalence and invertibility. J. Integral Equ. Applic. **21**, 499–543 (2009)
7. O. Chkadua, S.E. Mikhailov, D. Natroshvili, Analysis of direct segregated boundary-domain integral equations for variable-coefficient mixed BVPs in exterior domains. Anal. Applic.**11**, 1350006 (2013)
8. T.T. Dufera, S.E. Mikhailov, Boundary-domain integral equations for variable coefficient Dirichlet BVP in 2D unbounded domain. Anal. Probab. Applic. Comput. **1**, 481–492 (2019)
9. C. Fresneda-Portillo, S.E. Mikhailov, Analysis of boundary-domain integral equations to the mixed BVP for a compressible stokes system with variable viscosity. Commun. Pure Appl. Anal. **18**, 3059–3088 (2019)
10. G.C. Hsiao, W.L. Wendland, Boundary integral equations, in *Analysis Probability Applications and Computation* (Springer, Berlin/Heidelberg, 2008)
11. O.A. Ladyzhenskaya, The mathematical theory of viscous incompressible flow, in *Communications on Pure and Applied Analysis* (Gordon and Breach Science, New York, 1969)
12. W. McLean, L. Wolfgang, *Strongly Elliptic Systems and Boundary Integral Equations* (Cambridge University Press, Cambridge, 2000)
13. S. Mikhailov, C.F. Portillo, Boundary-domain integral equations equivalent to an exterior mixed BVP for the variable-viscosity compressible Stokes PDEs. Commun. Pure Appl. Anal. **20**, 1103–1133 (2021)
14. C. Miranda, Partial differential equations of elliptic type, in *Communications on Pure and Applied Analysis* (Springer, Berlin/Heidelberg, 1970)
15. A. Pomp, *The Boundary-Domain Integral Method for Elliptic Systems: With Application to Shells* (Springer Science & Business Media, Berlin/Heidelberg, 1998)
16. O. Steinbach, *Numerical Approximation Methods for Elliptic Boundary Value Problems: Finite and Boundary Elements* (Springer Science & Business Media, New York, NY, 2007)
17. R. Temam, *Navier-Stokes Equations: Theory and Numerical Analysis* (American Mathematical Society, Providence, 2001)

The Mixed Initial-Boundary Value Problem for Degenerate Hyperbolic Equation

Nurbek Kakharman

Abstract A new class of initial-boundary value problems with nonlocal conditions on the Newton potential for the case of a degenerate equation were set and studied.

1 Introduction

A number of studies have been conducted on the mixed Cauchy problem for noncharacteristically degenerate second-order hyperbolic equations, beginning with the work of M. L. Krasnov [1]. This was later extended to general degenerate higher-order equations [2–4]. The research into boundary value problems for mixed type equations, initiated by Tricomi [5], resulted in the exploration of novel boundary value problems for hyperbolic equations in the characteristic cone, which was first explored in [6–11].

In the study of the mixed Cauchy problem in a cylindrical domain, the lateral boundary conditions are usually local boundary conditions of Dirichlet-type boundary conditions.

It is well-known that the Cauchy problem is well posed for strictly hyperbolic equations. The situation is complicated for degenerate hyperbolic equations if the hyperbolic equation is degenerated along the characteristic line or if the coefficients of the hyperbolic equation at lower terms are singular. Therefore, it is natural to consider a "modified" Cauchy problem when the initial conditions are given with weight functions.

N. Kakharman (✉)
Institute of Mathematics and Mathematical Modeling, Almaty, Kazakhstan
e-mail: n.kakharman@math.kz

M. Ruzhansky, B. Torebek (eds.), *Extended Abstracts MWCAPDE 2023*,
Research Perspectives Ghent Analysis and PDE Center 1,
https://doi.org/10.1007/978-3-031-41665-1_16

2 Preliminaries

In [12], the Newton potential (volume potential) is given by a self-adjoint integral operator

$$u(x) = \int_{\Omega} \varepsilon(x, \xi)\rho(\xi)d\xi, \tag{2.1}$$

where $\rho(\xi) \in L_2(\Omega)$ and $\varepsilon(x, \xi)$ is a fundamental solution of the Laplace equation

$$-\Delta_x \varepsilon(x, \xi) = \delta(x - \xi), \quad \varepsilon(x, \xi) = \varepsilon(\xi, x), \tag{2.2}$$

the function $u \in W_2^2(\Omega)$ satisfies equation

$$-\Delta_x u = \rho(x) \tag{2.3}$$

and the lateral boundary condition

$$-\frac{u(x)}{2} + \int_{\partial\Omega} \left(\frac{\partial \varepsilon}{\partial n_\xi}(x, \xi) \cdot u(\xi) - \varepsilon(x, \xi) \cdot \frac{\partial u}{\partial n_\xi}(\xi) \right) d\xi = 0, \quad x \in \partial\Omega. \tag{2.4}$$

On the contrary, if $u \in W_2^2(\Omega)$ satisfies equation (2.3) and the boundary condition (2.4), then $u(x)$ coincides with the Newton potential (2.1).

The aim of this paper is to study the mixed Cauchy problem with condition (2.4).

3 Main Results

Let $\Omega \subset \mathbb{R}^n$ be a finite domain with a smooth boundary $\partial\Omega \subset C^2$, $D = \Omega \times [0, T]$ a cylindrical domain. Let $0 \leq \beta < 1$ and $f \in L_2(D)$.

In D, find a solution to the following equation:

$$Lu = \frac{\partial}{\partial t}\left(t^\beta \frac{\partial}{\partial t} u \right) - \Delta_x u = f(x, t), \tag{3.1}$$

which satisfies the initial conditions

$$u\,(x, 0) = 0, \qquad \lim_{t \to 0} t^\beta u_t(x, t) = 0, \tag{3.2}$$

and the lateral boundary condition

$$N[u] \equiv -\frac{u(x,t)}{2} + \int\limits_{\partial\Omega} \left(\frac{\partial \varepsilon}{\partial n_\xi}(x,\xi) \cdot u(\xi,t) - \varepsilon(x,\xi) \cdot \frac{\partial u}{\partial n_\xi}(\xi,t) \right) d\xi = 0,$$

$$t \in [0,T], \quad x \in \partial\Omega. \tag{3.3}$$

where $\varepsilon(x,\xi)$ is the fundamental solution of Laplace equation (2.2).

Given the complexity of the potential boundary condition (3.3), we will use the spectral decomposition method to obtain the necessary estimates for problem (3.1)–(3.3).

The eigenfunctions of the Newton potential satisfy the following equation [12]:

$$-\Delta e_m(x) = \lambda_m e_m(x), \tag{3.4}$$

and the boundary condition

$$N[e_m] \equiv -\frac{e_m(x)}{2} + \int\limits_{\partial\Omega} \left(\frac{\partial \varepsilon}{\partial n_\xi}(x,\xi) e_m(\xi) - \varepsilon(x,\xi) \frac{\partial e_m}{\partial n_\xi}(\xi) \right) d\xi = 0, x \in \partial\Omega. \tag{3.5}$$

Let $\{e_m(x)\}$ be a complete orthonormal system of eigenvectors of problem (3.4)–(3.5) [12](see also [13]); then the solution of (3.1)–(3.3) can be written in the form

$$u(x,t) = \sum_{|m|=1}^{\infty} u_m(t) e_m(x), \quad f(x,t) = \sum_{|m|=1}^{\infty} f_m(t) e_m(x), \tag{3.6}$$

where $e_m = e_{m_1 m_2 \ldots m_n}$ and

$$u_m(t) = \int\limits_{\Omega} u(x,t) e_m(x) dx, \quad f_m(t) = \int\limits_{\Omega} f(x,t) e_m(x) dx.$$

Substituting (3.6) into equation (3.1), for $u_m(t)$ we get the following one-dimensional Cauchy problem:

$$L_{K_m} u_m = \frac{d}{dt}\left(t^\beta \frac{d}{dt} u_m(t) \right) + \lambda_m u_m(t) = f_m(t), \tag{3.7}$$

$$u_m(0) = 0, \qquad \lim_{t \to 0} t^\beta \frac{\partial u_m}{\partial t}(t) = 0. \tag{3.8}$$

The general solution for the homogeneous equation $L_{K_m}u = 0$ has the following form [14]:

$$u_m(t) = c_1 t^{\frac{1-\beta}{2}} J_\nu\left(\frac{2\sqrt{\lambda_m}}{2-\beta} t^{\frac{2-\beta}{2}}\right) + c_2 t^{\frac{1-\beta}{2}} Y_\nu\left(\frac{2\sqrt{\lambda_m}}{2-\beta} t^{\frac{2-\beta}{2}}\right), \tag{3.9}$$

where $J_\nu(\cdot), Y_\nu(\cdot)$ are standard Bessel functions of the first and second kinds, respectively, [14], $\nu = \frac{1-\beta}{2-\beta}$, and c_1, c_2 is an arbitrary constant. It is straightforward to check that the solution for equation (3.7)–(3.8) is

$$\begin{aligned} u_m(t) =& \frac{\pi}{2-\beta} t^{\frac{1-\beta}{2}} Y_\nu\left(\frac{2\sqrt{\lambda_m}}{2-\beta} t^{\frac{2-\beta}{2}}\right) \int\limits_0^t f_m(\xi)\xi^{\frac{1-\beta}{2}} J_\nu\left(\frac{2\sqrt{\lambda_m}}{2-\beta}\xi^{\frac{2-\beta}{2}}\right) d\xi \\ &- \frac{\pi}{2-\beta} t^{\frac{1-\beta}{2}} J_\nu\left(\frac{2\sqrt{\lambda_m}}{2-\beta} t^{\frac{2-\beta}{2}}\right) \int\limits_0^t f_m(\xi)\xi^{\frac{1-\beta}{2}} Y_\nu\left(\frac{2\sqrt{\lambda_m}}{2-\beta}\xi^{\frac{2-\beta}{2}}\right) d\xi. \end{aligned} \tag{3.10}$$

Let

$$W^2_{2,t^\beta}(0,T) = \{v : v \in L_2(0,T) \text{ and } t^\beta \frac{\partial}{\partial t} v \in W^1_2(0,T),\ 0 \le \beta < 1\}$$

be a weighted Sobolev space equipped with the following norm:

$$\|v\|_{W^2_{2,t^\beta}(0,T)} := \left\| t^\beta \frac{\partial}{\partial t} v \right\|_{W^1_2(0,T)} + \|v\|_{L_2(0,T)}.$$

Lemma 3.1 *The solution $u_m \in W^{2,2}_{2,t^\beta}(0,T)$ of the one-dimensional problem* (3.7)–(3.8) *satisfies the following inequality:*

$$\left\| t^\beta \frac{\partial}{\partial t} u_m \right\|_{W^1_2(0,T)} + \|u_m\|_{L_2(0,T)} \le c\|f\|_{L_2(0,T)},$$

where constant c depends on β and λ_m.

Proof In the case $|\nu| < \frac{1}{2}$ and $t > 0$, for Bessel functions [15, p.360], we have

$$\begin{aligned} \left| t^{\frac{1-\beta}{2}} J_\nu\left(\frac{2\sqrt{\lambda_m}}{2-\beta} t^{\frac{2-\beta}{2}}\right)\right| &= \left| -t^{\frac{2-\beta}{2}} \frac{2^{1+\nu}(\sqrt{\lambda_m})^{-\nu} t^{-\frac{2-\beta}{2}}}{\pi^{\frac{1}{2}}\Gamma(\frac{1}{2}-\nu)} \int_1^\infty \frac{sin(t\xi)d\xi}{(\xi^2-1)^{\nu+\frac{1}{2}}} \right| \\ &\le c_0 \cdot (\sqrt{\lambda_m})^{-\nu}, \end{aligned} \tag{3.11}$$

and

$$\left|t^{\frac{1-\beta}{2}} Y_\nu\left(\frac{2\sqrt{\lambda_m}}{2-\beta} t^{\frac{2-\beta}{2}}\right)\right| = \left|-t^{\frac{2-\beta}{2}} \frac{2^{1+\nu}(\sqrt{\lambda_m})^{-\nu} t^{-\frac{2-\beta}{2}}}{\pi^{\frac{1}{2}}\Gamma(\frac{1}{2}-\nu)} \int_1^\infty \frac{cos(t\xi)d\xi}{(\xi^2-1)^{\nu+\frac{1}{2}}}\right|$$
$$\leq c_1 \cdot (\sqrt{\lambda_m})^{-\nu}. \tag{3.12}$$

The proof of lemma 3.1 follows by a straightforward calculation using inequalities (3.11) and (3.12). □

Let $\alpha > 0$, by $(-\Delta_x)^\alpha$ we denote the operator acting on $g \in L_2(D)$ by the formula

$$(-\Delta_x)^\alpha g = \sum_{|m|=1}^{\infty} g_m(t)\lambda_m^\alpha e_m(x),$$

$$\left\|(-\Delta_x)^\alpha g\right\|^2_{L_2(\Omega)}(t) = \sum_{|m|=1}^{\infty} \left|g_m(t)\lambda_m^\alpha\right|^2 < \infty, \quad t \in [0, T].$$

Let

$$W^{2,2}_{2,t^\beta}(D) = \{v : v \in L_2(D), \ \ \Delta_x v \in L_2(D) \ \text{ and } \ t^\beta \frac{\partial}{\partial t} v \in W^1_2(D), \ 0 \leq \beta < 1\}$$

be a weighted Sobolev space equipped with the following norm:

$$\|v\|_{W^{2,2}_{2,t^\beta}(D)} := \left\|t^\beta \frac{\partial}{\partial t} v\right\|_{W^1_2(D)} + \|\Delta_x v\|_{L_2(D)} + \|v\|_{L_2(D)}.$$

Theorem 3.2 *Let* $0 \leq \beta < 1$, $f \in L_2(D)$ *and*

$$\sum_{|m|=1}^{\infty} \left|(\lambda_m)^{1-\nu} f_m(\xi)\right|^2 < \infty.$$

Then there exists a unique solution $u \in W^{2,2}_{2,t^\beta}(D)$ *of the problem* (3.1)–(3.3) *that satisfies the following inequality:*

$$\left\|t^\beta \frac{\partial}{\partial t} u\right\|_{W^1_2(D)} + \|\Delta_x u\|_{L_2(D)} + \|u\|_{L_2(D)} \leq c\|f\|_{L_2(D)} + c_0 \left\|(\Delta_x)^{1-\nu} f\right\|_{L_2(D)},$$

where constants c *and* c_0 *depend only on* β.

Acknowledgments This research was funded by the Science Committee of the Ministry of Science and Higher Education of the Republic of Kazakhstan (Grant No. AP14971679).

References

1. M.L. Krasnov, Mixed boundary problems for degenerate linear hyperbolic differential equations second order. Mat. Sb. **91**(1), 29–84 (1959)
2. T.D. Dzhuraev, *Boundary Value Problems for Equations of Mixed and Mixed-Composite Types* (Fan, Tashkent, 1979)
3. V.N. Vragov, *Boundary Value Problems for Nonclassical Equations of Mathematical Physics* (Novosibirski, NGU, 1983). (in Russian)
4. A.I. Kozhanov, Linear inverse problems for a class of degenerate equations of Sobolev type. Vestnik Yuzhno-Ural'skogo Universiteta, Seriya Matematicheskoe Modelirovanie i Programmirovanie **11**, 33–42 (2012)
5. F. Tricomi, *On Second Order Linear Partial Differential Equation of Mixed Type* (Gostekhizdat, Moscow-Leningrad, 1947). (in Russian)
6. S. Gellerstedt, *Sur un problème aux limites pour uneéquation lineare aux dérivées partielles du second order de type mixte*. These pour le doctorat, Uppsala, 1935
7. A.V. Bitsadze, *Some Classes of Partial Differential Equations* (Science, Moscow, 1981)
8. A.M. Nakhushev, *Problems with Shift for Partial Differential Equations* (Science, Moscow, 2006)
9. T.S. Kal'menov, A criterion for the continuity of the solution of the Goursat problem for a certain degenerate equation. Differentsial'nye Uravneniya **8**(1), 41–54 (1972)
10. T.S. Kal'menov, The characteristic Cauchy problem for a certain class of degenerate hyperbolic equations. Differentsial'nye Uravneniya **9**(1), 84–96 (1973)
11. T.S. Kalmenov, D. Suragan, Initial-boundary value problems for the wave equation. Electr. J. Diff. Equ. **2014**(48), 1–6 (2014)
12. T.S. Kal'menov, D. Suragan, To spectral problems for the volume potential. Dokl. Math. **80**(2), 646–649 (2009)
13. N. Kakharman, T. Kal'menov, Mixed Cauchy problem with lateral boundary condition for noncharacteristic degenerate hyperbolic equations. Bound. Value Prob. **1**, 1–11 (2022)
14. E. Kamke, *Handbook of Ordinary Differential Equations* (Moscow, Nauka Publ., 1971). (in Russian)
15. M. Abramowitz, I. Stegun, *Handbook of Mathematical Functions* (Dover Publications Inc., New York, 1964)

A Criterion for Minimality of the Mixed Type Operators

Tynysbek Sh. Kalmenov

Abstract This work is devoted to find a necessary and sufficient condition for the solvability of the overdetermined problem for the mixed-type equation with Cauchy data on the entire boundary. The proof of the main result is based on the representation of the boundary conditions of the Newton (volume) potential of the multidimensional Laplace equation, which was obtained in the work (Kal'menov and Suragan, Dokl Math 80(2):646–649, 2009).

2000 Mathematics Subject Classification Primary 35J05; Secondary 31B10

1 Introduction

The study of local properties of a differential equation is usually tested on finitely differentiable functions, which serves as one of the main examples of an overdetermined boundary value problem. When finding solutions to a differential equation on the whole space, the properties of solutions on finite functions are also essentially used. On the other hand, when studying inverse problems, we have to solve the problem with overdetermined boundary conditions. Therefore, the study of problems with overdetermined boundary conditions for a differential equation is an actual problem.

Let $\Omega \subset R^n$ be a bounded domain with a smooth boundary $\partial\Omega$. Let us consider the following differential equation:

$$Lu \equiv \sum_{|\alpha|\leq m} a_\alpha(x)D^\alpha u(x) = f(x), \tag{1.1}$$

T. S. Kalmenov (✉)
Institute of Mathematics and Mathematical Modeling, Almaty, Kazakhstan
e-mail: kalmenov.t@mail.ru

M. Ruzhansky, B. Torebek (eds.), *Extended Abstracts MWCAPDE 2023*,
Research Perspectives Ghent Analysis and PDE Center 1,
https://doi.org/10.1007/978-3-031-41665-1_17

and its formally adjoint equation

$$L^{+}v \equiv \sum_{|\alpha|\leq m} (-1)^{|\alpha|} D^{\alpha} \left(a_{\alpha}(x)v(x)\right) = g(x). \tag{1.2}$$

A minimal operator L_0 is the closure of the differential operator L on subset of the functions $u \in C^{m+\alpha}(\overline{\Omega})$, with $\|D^{\alpha}u\|_{x\in\partial\Omega} = 0$, $|\alpha| = 0, 1, \ldots, m-1$, by $(L_0^{+})^{*}$; we denote the adjoint operator of the operator L_0 in the space $L_2(\Omega)$.

We recall the operator L a regular extension of the minimal operator L_0, if L is invertible in all $L_2(\Omega)$ and $L_0 \subset L \subset (L_0^{+})^{*}$, $\| L^{-1} \| < \infty$.

Let L_Q be a fixed differential operator generated by Eq. (1.1) and the regular boundary conditions.Then there is a linear bounded operator $K : \ker(L_0^{+})^{*} \rightarrow \ker(L_0^{+})^{*}$, such that $u \in D(L)$ is given as

$$u = u_0 + Kv + L_Q^{-1}v, \quad u_0 \in D(L_0), \quad v \in \ker L_0^{*}, \tag{1.3}$$

$$Lu = (L_0^{+})^{*}u = f. \tag{1.4}$$

Conversely, for any $K : \ker(L_0^{+})^{*} \rightarrow \ker(L_0^{+})^{*}$, there is a regular extension L, where L^{-1} is given by (1.3).

The theory of regular extensions for elliptic equations was first constructed by M.I.Vishik [1]. Vishik's result was extended to the case of Banach spaces by M. Otelbaev [2]. Moreover, in [2] there was given the description of the correct restriction of the maximal operator L_0^{*}, which includes not only boundary value problems but also problems with internal boundary conditions. A more complete bibliography may be found in [3–5].

In the present paper, we have obtained a minimality criterion for the degenerate elliptic and hyperbolic Gellerstedt potentials. In this case, the boundary condition of the Newtonian (volume) potential constructed by the authors [6] plays a main role.

2 Main Results

Let Δ_0 be a minimal Laplace operator for the overdetermined Cauchy problem, i.e., for $f \in L_2(\Omega)$, $D(\Delta_0) = W_2^2(\Omega)$,

$$\Delta_0 u_0 = \Delta u_0 = f, \quad x \in \Omega, \tag{2.1}$$

$$u_0 \mid_{x\in\partial\Omega} = 0, \quad \frac{\partial u_0}{\partial n} \mid_{x\in\partial\Omega} = 0. \tag{2.2}$$

By $G(x, y)$ we denote the Green function of the Dirichlet problem. Since $u_0 |_{x \in \partial\Omega} = 0$, then the condition of invertibility of Δ_0 is the condition

$$\frac{\partial}{\partial n} \int_{\Omega} G(x, y) f(y) dx dy = 0, \quad x \in \partial\Omega. \tag{2.3}$$

Since the Green function $G(x, y)$ is explicitly constructed only for special domains, this condition is difficult to verify.

On the other hand, if we use the decomposition $L_2(\Omega) = R(\Delta_0) \oplus \ker \Delta_0^*$, then the second condition for the invertibility of the minimal operator is the fulfillment of the equalities $(f, v) = 0$ for all $v \in \ker \Delta_0^*$, which is also difficult to verify.

Now we get the criterion of invertibility Δ_0 according to our method. We define the Newton (volume) potential in the following form:

$$u(x) = \int_{\Omega} \varepsilon(x - \xi) f(\xi) d\xi. \tag{2.4}$$

In the following, we use the representations of $\varepsilon(x)$ given as

$$\begin{aligned} \varepsilon(x) &= \frac{1}{2\pi} \ln |x|, \quad x \in R^2, \\ \varepsilon(x) &= -\frac{1}{\omega_n} \frac{1}{|x|^{n-2}}, \quad x \in R^n, \quad n \geq 3, \end{aligned} \tag{2.5}$$

where ω_n is the surface area of unit sphere R^n.

Theorem 2.1 *The minimal operator Δ_0 (a Cauchy problem for the Laplace equation with Cauchy data on the entire boundary $\partial\Omega$) is invertible in $L_2(\Omega)$ if and only if the following condition holds:*

$$\left. \int_{\Omega} \varepsilon(x - y) f(y) dy \right|_{x \in \partial\Omega} = 0. \tag{2.6}$$

The proof of Theorem 2.2 is based on the boundary property of the Newton (volume) potential $u(x)$ given by the integral form (2.4), which was obtained in[6, see Theorem 1] (see also [7]), namely, the following.

Theorem 2.2 *For any $f \in L_2(\Omega)$ the Newton potential defined by formula* (2.4) *belongs to $u \in W_2^2(\Omega)$ and satisfies the following boundary condition:*

$$-\frac{u(x)}{2} + \int_{\partial\Omega} \left(\varepsilon(x - y) \frac{\partial u(y)}{\partial n_y} - \frac{\partial \varepsilon(x - y)}{\partial n_y} u(y) \right) dy = 0, \quad x \in \partial\Omega. \tag{2.7}$$

Conversely, if $u \in W_2^2(\Omega)$, *satisfies Eq.* (1.1) *and the boundary condition* (2.7), *then* u *coincides with the Newton potential.*

Since the fundamental solutions for many equations, in particular, for classical equations of mathematical physics, are written out explicitly, the minimality criterion given by formula (2.6) is more effective than previously known ones.

Let $\Omega \subset R^2$ be a bounded domain with a smooth boundary $\partial\Omega$ and $y > 0$. By L_0, we denote the closure of the following Gellerstedt operator

$$Lu \equiv y^m u_{xx} + u_{yy} = f(x, y) \tag{2.8}$$

on a subset of functions $u \in C^\infty(\Omega)$ with the conditions

$$u\mid_{\partial\Omega} = 0, \quad \frac{\partial u}{\partial n}\mid_{\partial\Omega} = 0. \tag{2.9}$$

We define the fundamental solution of Eq. (2.8) in the following form [8]:

$$\begin{aligned} &\varepsilon(x_0, y_0, x, y) = q_1(x, y, x_0, y_0) \\ &= k_1 \left(\frac{4}{m+2}\right)^{4\beta-2} (r_1^2)^{-\beta}(1-\sigma)^{1-2\beta} F(1-\beta, 1-\beta; 2-2\beta; 1-\sigma); \end{aligned} \tag{2.10}$$

where

$$r_1^2 = (x - x_0)^2 + \frac{4}{(m+2)^2}\left(y^{\frac{m+2}{2}} + y_0^{\frac{m+2}{2}}\right),$$

$$\beta = \frac{1}{2(m+2)}, \quad k_1 = \frac{\Gamma(\beta)}{\Gamma(1-\beta)\cdot\Gamma(2\beta)},$$

and $F(a; b; c; z)$ is a hypergeometric function.

Let

$$u^+ = L_0^{-1} f^+ = \int_\Omega \varepsilon(x, y, \xi, \eta) f^+(\xi) d\xi d\eta$$

be a Gellersted potential in Ω^+.

Using the methods of Kal'menov and Suragan, we will make sure that the Gellerstedt potential (2.10) for any $f \in L_2(\Omega)$ satisfies the following condition:

$$-\frac{u(x,y)}{2} + \int\limits_{\partial\Omega} \left(\frac{\partial \varepsilon}{\partial n_\xi}(x,y,\xi,\eta)u(\xi,\eta) - \varepsilon(x,y,\xi,\eta)\frac{\partial u}{\partial n_\xi} \right)$$
$$\times d\xi d\eta = 0, \quad (x,y) \in \partial\Omega \tag{2.11}$$

where $\frac{\partial}{\partial n_\xi} = y^m n_1 \frac{\partial}{\partial \xi} + n_2 \frac{\partial}{\partial \eta}$ is the conormal derivative and n_1, n_2 are direction cosines.

Theorem 2.3 *For any $f \in L_2(\Omega)$ the minimal Gellerstedt operator is invertible if and only if the following condition holds:*

$$\int\limits_{\Omega} \varepsilon(x,y,\xi,\eta) f(\xi,\eta) d\xi d\eta \bigg|_{(x,y)\in\partial\Omega} = 0.$$

Acknowledgments This research was funded by the Science Committee of the Ministry of Science and Higher Education of the Republic of Kazakhstan (Grant No. AP14871460).

References

1. M.I. Višik, On general boundary problems for elliptic differential equations. Am. Math. Soc. Transl. **24**, 107–172 (1963)
2. M.O. Otelbaev, B.K. Kokebaev, A.N. Shynybekov, To the theory of contraction and expansion of operators. I. Izv. AN KazSSR, Ser. Phys.-Mat. **5**, 24–27 (1982)
3. M.O. Otelbaev, B.K. Kokebaev, A.N. Shynybekov, To the theory of contraction and expansion of operators. II. Izv. AN KazSSR, Ser. Phys.-Mat. **1**, 24–27 (1983)
4. B.K. Kokebaev, M. Otelbaev, A.N. Shynybekov, Some questions of expansion and restriction of operators. Docl. Akad. Nauk SSSR. **271**(6), 1307–1310 (1983)
5. N. Kakharman, K. Tulenov, L. Zhumanova, On hyponormal and dissipative correct extensions and restrictions. Math. Methods Appl. Sci. **45**(16), 9049–9060 (2022)
6. T.S. Kal'menov, D. Suragan, To spectral problems for the volume potential. Dokl. Math. **80**(2), 646–649 (2009)
7. N. Kakharman, T. Kal'menov, Mixed Cauchy problem with lateral boundary condition for noncharacteristic degenerate hyperbolic equations. Bound. Value Prob. **2022**, 1–11 (2022)
8. S. Gellerstedt, Sur un probleme aux limites pour une equation lineaire aux derivees partielles du second ordre de type mixte, Ph.D. Thesis, Almqvist Wiksells, Uppsala, 1935

An Analogue of the d'Alembert Formula for the Gurtin-Pipkin Equation with Multipoint Boundary Conditions for Zero Initial Data

Baltabek Kanguzhin and Zhalgas Kaiyrbek

Abstract The paper presents the derivation of an analogue of the d'Alembert formula for the Gurtin-Pipkin equation with inhomogeneous multipoint conditions. As a consequence, it is possible to obtain a statement about the possibility or impossibility of reaching a predetermined final regime in a fixed time, when the average temperature at a finite number of points acts as a control.

1 Introduction

The classical theory of thermal conductivity of a homogeneous and isotropic rod is based on the equation

$$\frac{\partial u(x,t)}{\partial t} = \alpha \frac{\partial^2 u(x,t)}{\partial x^2},$$

where $u = u(x,t)$ is the absolute temperature. This equation is parabolic and has one unpleasant property: a thermal perturbation at one point is instantly felt at all other points of the rod. In other words, the rate of propagation of the perturbation is infinite. Work [1] developed a general theory of thermal conductivity for nonlinear materials with memory, a theory that relates to it finite propagation velocities. Section 7 of [1] provides a linearized theory suitable for infinitesimal temperature gradients. In this case, the heat flux q is related to the temperature gradient $g = \frac{\partial u}{\partial x}$ by a linearized constitutive relation:

$$q = -\int_0^\infty a(s)g(t-s)ds. \tag{1.1}$$

B. Kanguzhin (✉) · Z. Kaiyrbek
Institute of Mathematics and Mathematical Modeling, Al-Farabi Kazakh National University, Almaty, Kazakhstan
e-mail: kanbalta@mail.ru

M. Ruzhansky, B. Torebek (eds.), *Extended Abstracts MWCAPDE 2023*, Research Perspectives Ghent Analysis and PDE Center 1,
https://doi.org/10.1007/978-3-031-41665-1_18

The function $a(s)$ is called the heat flux relaxation function. In particular, in [1] it was proposed to replace the heat equation with the following equation:

$$\frac{\partial u(x,t)}{\partial t} = \int_0^t G(s)\frac{\partial^2 u(x,t-s)}{\partial x^2}ds, \quad t>0 \tag{1.2}$$

where $G(s) = a'(s)$ at $0 \leq s \leq t$. Equation (1.2) is often called the Gurtin-Pipkin equation. Equations similar to the Eq. (1.2) also arise in the study of the behavior of viscoelastic fluids. Note that Eq. (1.2), like the wave equation, propagates perturbations with a finite velocity [2, 3]. Let us start with the following problem:

$$\rho\frac{\partial \vartheta(x,t)}{\partial t} = \int_0^t G(s)\frac{\partial^2 u(x,t-s)}{\partial x^2}ds, \quad 0<t<T, \quad 0<x<b \tag{1.3}$$

Temperature $\vartheta(x,t)$ obeys Eq. (1.3). At $t=0$ the temperature of the rod $\vartheta(x,0)$ is set. At the ends of the rod $x=0$ and $x=b$, there are heat sources that maintain the temperature $\vartheta(0,t)=\mu(t)$ and $\vartheta(b,t)=\eta(t)$. What should be the boundary regimes $\mu(t)$ and $\eta(t)$, so that in a fixed time T the rod is heated to a given temperature distribution $\vartheta(x,T)$? Without asking for generality, we can assume that $\vartheta(x,0)\equiv 0$. What final temperature distributions $\vartheta(x,T)$ can be achieved? This problem is called the boundary control problem. Here, the boundary modes $\mu(t)$ and $\eta(t)$ come into play as a control. We now formulate a modification of the above boundary control problem. We fix a natural number N and numbers $\alpha_0,\alpha_1,\ldots,\alpha_{N-1}$ such that $\sum_{j=0}^{N-1}\alpha_j = 1$, $\alpha_0 \neq 0$. As $\mu(t)$ we take the average value of $\sum_{j=0}^{N-1}\alpha_j\vartheta(\frac{b}{N}j,t)$, and the boundary value $\eta(t)=\vartheta(b,t)$ has the same meaning. It is required to find the boundary control $\mu(t)$ and $\eta(t)$ which, in a fixed time T, takes the rod from the initial state $\vartheta(x,0)$ to the final state $\vartheta(x,T)$ [4]. The main result of this article is that for the Gurtin-Pipkin Eq. (1.3) with multipoint boundary conditions

$$\sum_{j=0}^{N-1}\alpha_j\vartheta(\frac{b}{N}j,t)=\mu(t), \quad \vartheta(b,t)=\mu(t), \quad t>0 \tag{1.4}$$

with zero initial data, it has an analogue of the d'Alembert formula, which is valid for the wave equation. Success depends on T and b. If T is much smaller than b, then it is impossible to obtain an arbitrary desired final distribution $\vartheta(x,T)$.That is, if T is small, then there will not be enough time to heat the rod to the desired degrees. When T is "comparable" to b, then one can obtain an arbitrary desired final distribution $\vartheta(x,T)$, and, moreover, there is a "single" boundary regime. The sense in which the uniqueness of the boundary regime is understood is specified in the following paragraphs.

2 On the Solvability of a Single Multipoint Problem

In this subsection, we consider the following multipoint boundary value problem:

$$y''(x) = \lambda^2 \cdot y(x), \quad 0 < x < b, \tag{2.1}$$

$$\sum_{j=0}^{N-1} y(\frac{b}{N}j) = F, \quad y(b) = K, \tag{2.2}$$

where F and K are arbitrary constant numbers.

The choice of natural number N and numbers $\alpha_0, \alpha_1, \ldots, \alpha_{N-1}$ is indicated in the introduction.

Lemma 1 *If λ is such that*

$$\sum_{j=0}^{N-1} \alpha_j sh(\lambda \frac{b}{N}(N-j)) \neq 0,$$

then, the multipoint boundary value problem (2.1)–(2.2) has a unique solution for any F and K.

Furthermore, the solution is written in the form

$$y(x) = K \frac{\sum_{j=0}^{N-1} \alpha_j sh\lambda[x - \frac{b}{N}j]}{\sum_{j=0}^{N} \alpha_j sh\lambda \frac{b}{N}(N-j)} + F \frac{sh\lambda(b-x)}{\sum_{j=0}^{N-1} \alpha_j sh\lambda \frac{b}{N}(N-j)}.$$

The validity of Lemma 1 follows from the classical theorems of the theory of linear differential equations with constant coefficients. For further purposes, it is convenient to introduce the notation

$$\Delta = 2 \sum_{j=0}^{N-1} \alpha_j sh(\lambda \frac{b}{N}(N-j)).$$

It is easy to transform the function $\Delta(\lambda)$ to the form

$$\Delta(\lambda) = e^{\lambda b} \cdot \sum_{j=0}^{N-1} \alpha_j e^{-\lambda \frac{b}{N} j}(1 - e^{-2\lambda \frac{b}{N}(N-j)}).$$

Let us denote it by $z = e^{-\lambda \frac{b}{N}}$. *Then we have the relation*

$$e^{-\lambda b}\Delta(\lambda) = \sum_{j=0}^{N-1} \alpha_j z^j (1 - z^{2(N-j)}) = (1 - z^2)(\alpha_0 z^{2N-2} + d(z)),$$

where $d(z)$ *is a polynomial of degree not higher than* $(2N - 3)$.
Since

$$\Delta(\lambda) = \sum_{j=0}^{N-1} \alpha_j (\frac{1}{z^{N-j}} - z^{N-j}),$$

then it is convenient to introduce a polynomial

$$P(z) = \sum_{j=0}^{N-1} \alpha_j (\frac{1}{z^{N-j}} - z^{N-j}).$$

If $P(a) = 0$, *then* $P(\frac{1}{a}) = 0$. *Therefore, the polynomial* $P(z)$ *can be decomposed into linear factors, that is,*

$$P(z) = \alpha_0 (1 - z)(1 + z) \prod_{j=1}^{N-1} (a_j - z)(\frac{1}{a_j} - z),$$

where the numbers a_j *at* $j = 1, 2, \ldots, N - 1$ *are chosen so that* $P(a_j) = 0$.

Therefore, for $\Delta(\lambda)$ *there is a decomposition*

$$\Delta(\lambda) = \alpha_0 (1 - e^{-\lambda \frac{b}{N}})(1 + e^{-\lambda \frac{b}{N}}) \prod_{j=1}^{N-1} (a_j - e^{-\lambda \frac{b}{N}})(\frac{1}{a_j} - e^{-\lambda \frac{b}{N}}).$$

In what follows, we assume that $a_k \neq a_j$ *at* $k \neq j$. *Then the expression* $\frac{1}{\Delta(\lambda)}$ *is decomposed into simple fractions*

$$\frac{1}{\Delta(\lambda)} = \frac{A_0^+}{1 - e^{-\lambda \frac{b}{N}}} + \frac{A_0^-}{1 + e^{-\lambda \frac{b}{N}}} + \sum_{j=1}^{N-1} \left(\frac{\frac{A_j^+}{a_j}}{1 - \frac{1}{a_j} e^{-\lambda \frac{b}{N}}} + \frac{A_j^- a_j}{1 - a_j e^{-\lambda \frac{b}{N}}} \right), \tag{2.3}$$

where

$$A_0^{\pm} = \lim_{z \to \mp 1} \frac{1 \mp z}{P(z)}, \quad A_j^{\pm} = \lim_{z \to a_j^{\pm 1}} \frac{a_j^{\pm 1} - z}{P(z)}. \tag{2.4}$$

Lemma 1 and decomposition (2.3) imply the assertion.

Lemma 2 *Let N be a fixed natural number. The numbers $\alpha_0, \alpha_1, \ldots, \alpha_{N-1}$ are chosen so that*

(i) $\sum_{j=0}^{N-1} \alpha_j = 1, \quad \alpha_0 \neq 0$;
(ii) Polynomial

$$P(z) = \sum_{j=0}^{N-1} \alpha_j \left(\frac{1}{z^{N-j}} - z^{N-j} \right)$$

has no multiple roots. Its roots not equal to ± 1 will be denoted by a_j, a_j^{-1} at $j = 1, 2, \ldots, N-1$.

If $e^{-\lambda \frac{b}{N}}$ is not equal to the numbers $\{+1, -1, a_j, a_j^{-1}, j \geq 1\}$, then the multipoint boundary value problem (2.1)–(2.2) has a unique solution, representable in the form

$$y(x) = 2(K \sum_{k=0}^{N-1} \alpha_k sh\lambda \left(x - \frac{b}{N} k \right) + F sh\lambda(b-x)) \times$$

$$\times \left[\frac{A_0^+}{1 - e^{-\lambda \frac{b}{N}}} + \frac{A_0^-}{1 + e^{-\lambda \frac{b}{N}}} + \sum_{j=1}^{N-1} \left(\frac{\frac{A_j^+}{a_j}}{1 - \frac{1}{a_j} e^{-\lambda \frac{b}{N}}} \right) + \frac{A_j^- a_j}{1 - a_j e^{-\lambda \frac{b}{N}}} \right].$$

Here the numbers $A_j^{\pm}$ are calculated by formulas (2.4).

3 An Analogue of the d'Alembert Formula for a Mixed Problem for the Gurtin-Pipkin Equation

In this subsection, we prove the unique solvability of the following mixed problem for the Gurtin-Pipkin equation:

$$\rho \frac{\partial \vartheta(x,t)}{\partial t} = \int_0^t G(t-s) \frac{\partial \vartheta(x,s)}{\partial x^2} ds, \quad 0 < x < b, \quad 0 < t, \tag{3.1}$$

with initial condition

$$\vartheta(x, 0) = 0 \tag{3.2}$$

and boundary conditions

$$\sum_{j=0}^{N-1} \alpha_j \vartheta(\frac{b}{N} j, t) = \mu(t), \quad \vartheta(b, t) = \mu(t), \quad t > 0. \tag{3.3}$$

Here and below, we assume that the relaxation function $G(t)$ is chosen in the form

$$G(t) = \sum_{k=0}^{\infty} c_k e^{-\lambda_k t},$$

where $c_k > 0, \lambda_k > 0, inf\{\lambda_k : k \geq 1\} = \delta >, \quad G(0) = \sum_{k=0}^{\infty} c_k < \infty$.

The density ρ is assumed to be a positive value throughout.

Denote by $\hat{f}(s)$ the Laplace transform of the function $f(t)$, i.e.,

$$\hat{f}(s) = \int_0^{\infty} e^{-st} \cdot f(t) dt.$$

Then applying the Laplace transform to problem (3.1)–(3.2)–(3.3), we obtain

$$\rho \cdot s \cdot \hat{\vartheta}(x, s) = \hat{G}(s) \cdot \frac{d^2 \hat{\vartheta}(x, s)}{dx^2}, \quad 0 < x < b, \tag{3.4}$$

$$\sum_{j=0}^{N-1} \alpha_j \hat{\vartheta}(\frac{b}{N} j, s) = \hat{\mu}(s), \quad \hat{\vartheta}(b, s) = \hat{\eta}(s). \tag{3.5}$$

Denote by $\lambda^2(s) = \frac{\rho s}{\hat{G}(s)}$. On the complex S plane, we choose a cut along the negative semi-axis and, by the monodromy theorem, calculate the single-valued branch $\lambda(s) = \sqrt{\frac{\rho s}{\hat{G}(s)}}$. Then the solution of the multipoint boundary value problem (3.4)–(3.5) according to Lemma 2 is written as follows:

$$\hat{\vartheta}(x, s) = 2 \left(\hat{\eta}(s) \sum_{k=0}^{N-1} \alpha_k \cdot sh\lambda(s)(x - \frac{b}{N} k) + \hat{\mu}(s) sh\lambda(s)(b - x) \right) \cdot \tag{3.6}$$

$$\cdot e^{-\lambda(s) b} \left[\frac{A_0^+}{1 - e^{-\lambda(s) \frac{b}{N}}} + \frac{A_0^-}{1 + e^{-\lambda(s) \frac{b}{N}}} \right.$$

$$\left. + \sum_{j=1}^{N-1} \left(\frac{\frac{A_j^+}{a_j}}{1 - \frac{1}{a_j} e^{-\lambda(s) \frac{b}{N}}} + \frac{A_j^- a_j}{1 - a_j e^{-\lambda(s) \frac{b}{N}}} \right) \right].$$

Since the relaxation function has the form $G(t) = \sum_{k=0}^{\infty} c_k e^{-\lambda_k t}$, then repeating the reasoning from [2], we get the following statement: If $Re\, s > 0$, then $Re\, \lambda(s) > 0$. You can also understand that $\lim_{Re\, s \to +\infty} Re\, \lambda(s) = +\infty$. Therefore, for sufficiently large $Re\, s > 0$, the expansions

$$\frac{1}{1 - \theta e^{-\lambda(s)\frac{b}{N}}} = 1 + \theta e^{-\lambda(s)\frac{b}{N}} + \theta^2 e^{-2\lambda(s)\frac{b}{N}} + \dots$$

for any finite constant θ. Then the right side of equality (3.6) for sufficiently large signs $Re\, s$ takes the form

$$\hat{\vartheta}(x,s) = \hat{\eta}(s)\left(\sum_{m=N}^{\infty} e^{-\lambda(s)(\frac{b}{N}m - x)}\varepsilon_m + \sum_{m=1}^{\infty} e^{-\lambda(s)(x+\frac{b}{N}m)}\delta_m\right) \tag{3.7}$$

$$+ \hat{\mu}(s)\left(\sum_{m=0}^{\infty} e^{-\lambda(s)(x+\frac{b}{N}m)}\gamma_m + \sum_{m=N}^{\infty} e^{-\lambda(s)(\frac{b}{N}m - x)}\beta_m\right),$$

where the numbers $\varepsilon_m, \delta_m, \gamma_m, \beta_m$ are some constant numbers.
Now it remains to calculate the inverse Laplace transform from the right side of the last equality (3.7). To do this, it suffices to calculate the intervals of the form

$$I_m^{\pm}(x,t) = \frac{1}{2\pi i}\int_{\gamma - i\infty}^{\gamma + i\infty} e^{ts} \cdot \frac{e^{-\lambda(s)(\frac{b}{N}m \pm x)}}{s} ds,$$

where γ a large enough real number.
In [2], the required integral $I_m^{\pm}(x,t)$ at $m = 0, 1, \dots,$ already calculated by the contour integration method. According to the results of [2], we have

$$I_m^{\pm}(x,t) = \begin{cases} 0, & at\ t < \alpha(\frac{b}{N}m \pm x), \\ f(\frac{b}{N}m \pm x, t), & at\ t \geq \alpha(\frac{b}{N}m \pm x). \end{cases}$$

Here, the quantity $\frac{1}{\alpha} = \frac{\sqrt{G(0)}}{\sqrt{\rho}}$ represents the final velocity of the thermal perturbation. We are not interested in the specific form of the function $f(x,t)$ in this article, although certain properties are indicated in [5].
Taking into account the above information about the integral $I_m^{\pm}(x,t)$, remembering the properties of the Laplace transform, we can write the solution of the mixed

problem (3.1)–(3.2)–(3.3) in the following form:

$$\vartheta(x,t) = \sum_{m=N}^{\infty} \varepsilon_m \int_{\alpha(\frac{b}{N}m-x)}^{t} \frac{d}{dt}\eta(t-\tau) f(\frac{b}{N}m - x, \tau) H(t - \alpha(\frac{b}{N}m - x)) d\tau$$
$$+ \sum_{m=1}^{\infty} \delta_m \int_{\alpha(\frac{b}{N}m+x)}^{t} \frac{d}{dt}\eta(t-\tau) f(\frac{b}{N}m + x, \tau) H(t - \alpha(\frac{b}{N}m + x)) d\tau \tag{3.8}$$
$$+ \sum_{m=1}^{\infty} \theta_m \int_{\alpha(\frac{b}{N}m+x)}^{t} \frac{d}{dt}\mu(t-\tau) f(\frac{b}{N}m + x, \tau) H(t - \alpha(\frac{b}{N}m + x)) d\tau$$
$$+ \sum_{m=1}^{\infty} \beta_m \int_{\alpha(\frac{b}{N}m-x)}^{t} \frac{d}{dt}\mu(t-\tau) f(\frac{b}{N}m - x, \tau) H(t - \alpha(\frac{b}{N}m - x)) d\tau.$$

Formula (3.8) is an analogue of the d'Alembert formula [5] for the wave equation

$$\frac{\partial^2}{\partial t^2} u(x,t) = a^2 \frac{\partial^2 u(x,t)}{\partial t^2}, \quad 0 < x < b, \ t > 0$$

with zero initial conditions

$$u(x,0) = 0, \quad \frac{\partial u(x,0)}{\partial t} = 0, \ 0 < x < b$$

and boundary conditions

$$u(0,t) = \mu(t), \ u(b,t) = \eta(t), \quad t > 0.$$

In this case, the role of the parameter a is played by the quantity $\frac{1}{a}$. To demonstrate the finite rate of propagation of perturbations, we give an example when $\eta(t) \equiv 0$. Then from formula (3.8), it follows

$\vartheta(x,t) \equiv 0$ at $t < \alpha x$,
$\vartheta(x,t) = \int_{\alpha x}^{t} \frac{d}{dt}\mu(t-\tau) f(x,\tau) d\tau$ at $\alpha x \le t < \alpha(\frac{b}{N} + x)$,

and $\vartheta(x,t) = \int_{\alpha x}^{t} \frac{d}{dt}\mu(t-\tau) f(x,\tau) d\tau + \int_{\alpha(\frac{b}{N}+x)}^{t} \frac{d}{dt}\mu(t-\tau) f(\frac{b}{N} + x, \tau) d\tau$ at $\alpha(\frac{b}{N} + x) \le t < \alpha(2\frac{b}{N} + x)$.

References

1. M.E. Gurtin, A.C. Pipkin, A general theory of heat conduction with finite wave speeds. Arch. Ration. Mech. Anal. **31**, 113–126 (1968)
2. A. Nerrain, D. Joseph, Linearized dynamics for step jumps of velocity and displacement of shearing flows of a simple fluid. Rheol. Acta **21**, 228–250 (1982)
3. M. Renardy, Some remarks on the propagation and non-propagation of discontinuities in lineary viscoelastic liquids. Rheol. Acta **21**, 250–254 (1982)
4. M. Slemrod, A hereditary partial differential equation with application in the theory of simple fluids. Arch. Ration. Mech. Anal. **62**, 303–321 (1976)
5. B.E. Kanguzhin, Propagation of nonsmooth waves under singular perturbations of the wave equation. Eurasian Math. J. **13:3**, 41–50 (2022)

On Fredholm Solvability and on the Index of the Generalized Neumann Problem for an Elliptic Equation of High Order on a Plane

Bakytbek Koshanov and Maksat Bakytbek

Abstract In this paper, we investigate the Fredholm solvability of the generalized Neumann problem for a high-order elliptic equation on a plane. The equivalence of the solvability conditions of the generalized Neumann problem to the complementarity condition (Shapiro-Lopatinskii condition) is proved. The formula for the index of the specified problem in the class $C^{2l,\mu}(\overline{D})$ is calculated.

1 Introduction

Complex analysis methods constitute a classical direction in the study of elliptic equations and equations of mixed type on the plane. At present, active research is being carried out in this direction in many mathematical centers of the world.

In a simply connected domain D on a plane bounded by a simple smooth contour $\Gamma \in C^{2l,\mu},\ l \geq 2,\ 0 < \mu < 1$, for an elliptic equation of the $2l$-order

$$\sum_{r=0}^{2l} a_r \frac{\partial^{2l} u}{\partial x^{2l-r} \partial y^r} + \sum_{0 \leq r \leq k \leq 2l-1} a_{rk}(x) \frac{\partial^k u}{\partial x^{k-r} \partial y^r} = F \tag{1.1}$$

B. Koshanov (✉)
International University of Information Technology, Al-Farabi Kazakh National University, Almaty, Kazakhstan
e-mail: koshanov@list.ru

M. Bakytbek
Kazakh Academy of Transport and Communication named after M.Tynyshbayev, Almaty, Kazakhstan
e-mail: 87759002818@mail.ru

M. Ruzhansky, B. Torebek (eds.), *Extended Abstracts MWCAPDE 2023*, Research Perspectives Ghent Analysis and PDE Center 1,
https://doi.org/10.1007/978-3-031-41665-1_19

with constant highest coefficients $a_r \in \mathbb{R}$ and lower coefficients $a_{rk} \in C^{\mu}(\overline{D})$, consider the boundary value problem

$$\left.\frac{\partial^{k_j-1}u}{\partial n^{k_j-1}}\right|_{\Gamma} = f_j, \quad j = 1, \ldots, l, \tag{1.2}$$

where $n = n_1 + in_2$ means the unit outward normal and natural k_j are subject to the condition $1 \leq k_1 < k_2 < \ldots < k_l \leq 2l$.
Here and below, the normal derivative $(\partial/\partial n)^k$ of order k is understood as the boundary operator

$$\left(n_1\frac{\partial}{\partial x} + n_2\frac{\partial}{\partial y}\right)^k = \sum_{r=0}^{k}\binom{k}{r}n_1^r n_2^{k-r}\frac{\partial^k}{\partial x^r \partial y^{k-r}}$$

and a similar meaning has the boundary operator $(\partial/\partial e)^k$ with respect to the unit tangent vector $e = e_1 + ie_2 = i(n_1 + in_2)$.

This problem becomes the Dirichlet problem if $k_j = j$ and becomes the Neumann problem if $k_j = j + 1$.

Materials and methods. The statement of this problem for $k_{j+1} - k_j \equiv 1$ for a polyharmonic equation originates from [1], where for $k_1 \geq 2$ it is called the generalized Neumann problem. This name is further retained for an arbitrary set of indicators k_j. Another variant of the Neumann problem, based on the variational principle, was proposed in [2]. If lower coefficients and the right-hand side are equal to zero, then problem (1.1), (1.2) was studied in [3, 4]. The case where they are different from zero, is studied detail in [5] in the space

$$C_a^{2l-1,\mu}(\overline{D}) \equiv \{u : u \in C^{2l}(D) \cap C^{2l-1,\mu}(\overline{D}), \quad \sum_{r=0}^{2l} a_r \frac{\partial^{2l}u}{\partial x^{2l-r}\partial y^r} \in C^{\mu}(\overline{D})\},$$

in particular, a necessary and sufficient condition for its Fredholm property is found.

Results and Discussion. The study of Eq. (1.1) in the multidimensional case is of great scientific interest. In the model case, Eq. (1.1) is called the polyharmonic equation $\Delta^l u(x) = F(x)$, $x = (x_1, \ldots, x_n) \in D \subseteq \mathbb{R}^n$. It is known that for this polyharmonic equation, the Dirichlet problem is uniquely solvable for any right-hand side of the equation. In [6, 7], a new representation of the Green function of the Dirichlet problem for a polyharmonic equation in a multidimensional sphere is constructed explicitly. In [8, 9], a representation of the Green function of the Neumann problem for the Poisson equation in a multidimensional unit ball is obtained. In [10–12], Green functions of Dirichlet, Neumann, and Robin problems for biharmonic and polyharmonic equations in a circle, semicircle, semi-ring, triangle, and other standard plane domains are constructed. The results of these works are based on the classical theory of integral representations for analytic, harmonic, and polyharmonic functions on the plane.

The paper [13] describes well-posed boundary value problems for a polyharmonic operator.

In this article, for a higher-order elliptic equation, it is proposed to develop a new functional-theoretical approach based on the use of functions that are analytic according to Douglis [14–16].

In the early 1960s, it became clear [17, 18] that in the theory of elliptic equations and systems, an important role is played by functions analytic in the sense of Douglis. These functions are solutions of a first-order elliptic system generalizing the classical Cauchy-Riemann system. In [19, 20], this approach has already been successfully applied to problems of the plane theory of elasticity (including the general anisotropic case). However, for domains with piecewise smooth boundaries and equations with continuous coefficients and, especially, for problems with nonlocal boundary conditions, this approach requires its further development.

In this paper, under the assumption $\Gamma \in C^{2l,\mu}$, the results obtained in [5] are extended to the more standard class $C^{2l,\mu}(\overline{D})$.

2 Description of the Fredholm Property of Problem (1.1),(1.2)

An operator $A \in \mathfrak{L}(X, Y)$ is called *Fredholm* if its kernel $Ker A$ and cokernel are finite-dimensional, and

$$\dim Im A = \dim (coker A)^{\perp}.$$

The Fredholm property and the index of a problem are understood in relation to its bounded operator, in our case

$$X \equiv C^{2l,\mu}(\overline{D}) \to Y \equiv C^{\mu}(\overline{D}) \times \prod_{j=1}^{l} C^{2l-k_j+1,\mu}(\Gamma).$$

Let $\nu_k,\ 1 \le k \le m$ be all different roots of the characteristic equation

$$\chi(z) = a_{2l} \prod\nolimits_{k=1}^{m} (z - \nu_k)^{l_k} \prod\nolimits_{k=1}^{m} (z - \overline{\nu_k})^{l_k} = 0,$$

in the upper half-plane and the l_k- multiplicity of the kth root, so that their total multiplicity $l_1 + \ldots + l_m$ is equal to l. The ellipticity condition is that $a_{2l} \neq 0$ and the roots of the characteristic polynomial $\chi(z) = a_0 + a_1 z + \ldots + a_{2l} z^{2l}$ do not lie on the real axis.

Let us introduce functions that are fractionally linear in z

$$\omega(e, z) = \frac{e_2 - e_1 z}{e_1 + e_2 z} = \frac{n_1 + n_2 z}{e_1 + e_2 z}, \tag{2.1}$$

where the dependence on the unit tangent vector $e = e_1 + ie_2$ to the contour Γ is indicated explicitly. For definiteness, the vector e is oriented positively with respect to the domain D, i.e., D lies to the left of this vector.

For an analytic l-vector-function $g(z) = (g_1(z), \ldots, g_n(z))$, in a neighborhood of points $z_1, \ldots, z_m$, we introduce the block $l \times l$-matrix

$$W_g(z_1, \ldots, z_m) = (W_g(z_1), \ldots, W_g(z_m)), \tag{2.2}$$

where matrix $W_g(z_k) \in \mathbb{C}^{l \times l_k}$ is composed of column vectors

$$g(z_k), g'(z_k), \ldots, \frac{1}{(l_k - 1)!} g^{(l_k - 1)}(z_k).$$

As g below we use a vector with components

$$g_j(z) = z^{k_j - 1}, \ 1 \le j \le l. \tag{2.3}$$

In this notation, the following theorem was proved in [5].

Theorem 2.1

(a) *Problem (1.1), (1.2) is Fredholm if and only if*

$$\det W_g[\omega(e, \nu_1), \ldots, \omega(e, \nu_m)] \neq 0, \quad e \in \mathbb{T}, \tag{2.4}$$

where $\mathbb{T}$ stands for the unit circle. Accordingly, the index of this problem is given by the formula

$$æ = -2 \left[\frac{1}{2\pi} \arg \det W_g[\omega(e, \nu_1), \ldots \omega(e, \nu_m)] \Big|_{\mathbb{T}} + l^2 - \sum\nolimits_{j=1}^{m} l_j^2 \right], \tag{2.5}$$

where the increment is taken along the counterclockwise unit circle.

(b) *In each of the following two cases:*

$$k_j = k_1 + j - 1, \ 1 \le j \le l; \tag{2.6}$$

$$m = 1, \ \nu_1 = \nu, \tag{2.7}$$

the problem (1.1), (1.2) is Fredholm, and its index is zero.

Obviously, condition (2.4) depends only on the set of numbers $k_1, k_2, \ldots, k_l$, so that for fixed k_j, when it is satisfied, problem (1.1), (1.2) is Fredholm in any domain.

3 Equivalence of the Fredholm Property Condition for Problem (1.1), (1.2) to the Complementarity Condition

From the point of view of the general elliptic theory [21], problem (1.1), (1.2) is Fredholm in the space $C^{2l,\mu}(\overline{D})$ if and only if its boundary conditions satisfy as follows called the complementarity condition (or the Shapiro-Lopatinskii condition [22]). In this case, [23] also says that the boundary conditions (1.2) cover the differential operator

$$L = \sum_{r=0}^{2l} a_r \frac{\partial^{2l}}{\partial x^{2l-r} \partial y^r},$$

corresponding to the main part (1.1). The indicated condition is as follows: starting from the fixed point $t \in \Gamma$ differentiation with respect to x and y in the expressions of the operators L and B_j, we replace, respectively, $e_1(t)+zn_1(t)$ and $e_2(t)+zn_2(t)$. As a result, we get polynomials

$$L(n, z) = \sum_{r=0}^{2l} a_r (e_1 + zn_1)^{2l-r} (e_2 + zn_2)^r$$

and

$$B_j(z) = [n_1(e_1 + zn_1) + n_2(e_2 + zn_2)]^{k_j - 1} = z^{k_j - 1}, \quad 1 \leq j \leq l.$$

Since $n_1 = e_2, \; n_2 = -e_1$, in the notation (2.1) the polynomial $L(n, z)$ can be written in the form

$$L(n, z) = (e_1 + zn_1)^{2l} \sum_{r=0}^{2l} a_r [-\omega(z)]^r,$$

so $L(z) = 0$ is equivalent to

$$-\omega(z) = \nu, \tag{3.1}$$

where ν is an arbitrary root of the characteristic equation $\chi(z)$. Moreover, their corresponding multiplicities coincide.

Obviously, transformation (2.1) takes the upper half-plane onto itself, so the transformation $z \to -\omega(\bar{z})$ also has a similar property. In particular, the polynomial l-degree

$$L^+(z) = (z - z_1)^{l_1} \dots (z - z_m)^{l_m}, \quad -\omega(z_j) = \bar{\nu}_j, \tag{3.2}$$

is formed by the roots of the equation $L(n, z) = 0$ lying in the upper half-plane.

In the adopted notation, the **complementarity condition** consists in the linear independence of the polynomials $B_j(z)$, $1 \le j \le l$, modulo the polynomial $L^+(z)$. Thus, this condition should be equivalent to condition (2.4) obtained in another way. This fact can be established directly.

Lemma 3.1 *Condition (2.4) is satisfied if and only if the polynomials* $B_j(z) = z^{k_j-1}$, $1 \le j \le l$, *are linearly independent into the modulus of the polynomial* $L^+(z)$.

Proof Suppose that these polynomials are linearly dependent modulo $L^+(z)$, that is, there is a nontrivial linear combination of them $B = \alpha_1 B_1 + \ldots + \alpha_l B_l$, a multiple of L^+. In the notation (2.3), the polynomial $B_j = g_j$, so this fact can be written in the form

$$B(z) = \sum_{j=1}^{l} \alpha_j z^{k_j-1} = Q(z)L^+(z)$$

with some polynomial Q. In accordance with (10), this relation means that the polynomial B at the points z_k has a zero of order l_k or, which is equivalent,

$$\sum_{j=1}^{l} \alpha_j g_j^{(s)}(z_k) = 0, \quad 0 \le s \le l_k - 1, \ 1 \le k \le m. \tag{3.3}$$

These equalities represent a homogeneous system of l equations for $\alpha_1, \ldots, \alpha_l$. It can be seen from definition (2.2) that the matrix of this system coincides with the matrix transposed to $W_g(z_1, \ldots, z_m)$. Therefore, a nonzero solution to system (3.3) is possible if and only if

$$\det W_g(z_1, \ldots, z_m) = 0. \tag{3.4}$$

According to definition (2.1), equality (3.1) is equivalent to $\overline{z}_j = \omega(\nu_j)$; therefore, equality (3.4) can be expressed in the form of vanishing of the determinant on the left-hand side of (2.4). Thus, violation of the complementarity condition is equivalent to violation of condition (2.4), which completes the proof of the lemma. □

4 Continuation of the Description of the Fredholm Property Condition for Problem (1.1), (1.2)

Note that formulas (2.4) and (2.5) will not change if we go from the vector g to the vector q defined by the relation

$$g(z) = z^{k_1-1} q(z), \quad q(z) = (1, z^{s_1}, \ldots, z^{s_{l-1}}), \quad s_j = k_{j+1} - k_1. \tag{4.1}$$

Moreover, as noted in [3, 5], the determinants of the matrices W with these vectors are related by the relation

$$\det W_g(z_1, \ldots, z_m) = \prod_j z_j^{l_j(l_1-1)} W_q(z_1, \ldots, z_m).$$

In these designations, condition (2.4) and the index formula (2.5) can be given a different form, more convenient for use. Let us introduce the fractional-linear functions

$$\gamma_k(z) = \frac{\nu_k - z}{1 + \nu_k z}, \qquad 1 \le k \le m, \tag{4.2}$$

and the function

$$R(z) = \det W_g(\gamma_1(z), \ldots, \gamma_m(z)). \tag{4.3}$$

This transformation $\gamma_k(z)$ swaps the points $\pm i$ and is involutive

$$\gamma(\pm i) = \mp i, \quad \gamma[\gamma(z)] \equiv z. \tag{4.4}$$

Moreover, for $\nu_k = i$, the identity $\gamma_k(z) \equiv i$ holds.

Thus, the rational function $R(z)$ admits poles only at the points $\nu_k' = -1/i \neq i$ in the upper half-plane and, in particular, is analytic in the lower half-plane. In addition, for $m \ge 2$ ($m \ge 3$) at the points $\zeta = -i$ ($\zeta = \pm i$) vanishes. This follows from (4.2) and the fact that for $z_1 = \ldots = z_m = -i$ ($z_1 = \ldots = z_m = \pm i$) matrix (2.2) has the same columns, so its determinant is zero. In particular, the function $R(z)$ is completely divisible by $z^2 + 1$.

Theorem 4.1 *Problem (1.1), (1.2) is Fredholm if and only if the rational function $R(\zeta)$ has no real roots on the extended real line $\overline{\mathbb{R}} = \mathbb{R} \cup \{\infty\}$, and under this condition, its index is given by the formula*

$$æ = 4(n - \sum\nolimits_{i<j} l_i l_j), \tag{4.5}$$

where n is the number of zeros of this function in the lower half-plane of the function, taking into account their multiplicity.

Proof Comparison of definitions (2.2) and (4.2) implies that

$$\det W_g[\omega(e, \nu_1), \ldots, \omega(e, \nu_m)] = R(e_2/e_1).$$

The function $\omega(e, \nu)$ in (2.1) is even in the variable $e \in \mathbb{T}$ and therefore the quantity

$$\arg \det W_g[\omega(e, \nu_1), \ldots \omega(e, \nu_m)]\Big|_{\mathbb{T}} = 2\arg \det W_g[\omega(e, \nu_1), \ldots \omega(e, \nu_m)]\Big|_{\mathbb{T}^+},$$

where $\mathbb{T}^+$ is a semicircle in the right half-plane. The mapping $e = e_1 + ie_2 \to t = e_2/e_1$ realizes a homeomorphism of this semicircle onto the extended real line $\overline{\mathbb{R}}$, and bypassing it from the point $e = -i$ to $e = i$ corresponds to movement on a straight line in the positive direction. Therefore, condition (2.4) is equivalent to the fact that the function R has no real roots on the extended real line, and the equality

$$\arg \det W_g[\omega(e, \nu_1), \dots \omega(e, \nu_m)]\Big|_{\mathbb{T}} = 2\arg \det W_g[\gamma_1(t), \dots \gamma_m(t)]\Big|_{-\infty}^{+\infty}.$$

As a result, formula (2.5) becomes

$$æ = -\frac{2}{\pi}\arg R(t)\Big|_{-\infty}^{+\infty} - 2(l^2 - \sum\nolimits_{j=1}^{m} l_j^2).$$

The rational function R has no poles in the lower half-plane, so, taking into account Rouche's theorem, the previous equality coincides with (4.4). It is only necessary to take into account that the lower half-plane remains on the left when traversing the straight line in the negative direction and that

$$l^2 - \sum\nolimits_{j=1}^{m} l_j^2 = 2\sum_{i<j} l_i l_j.$$

The theorem is proved. □

Let us consider in more detail the function $\gamma(z)$, defined by (4.2) with $\nu = \nu_k$. For $\nu = i$ this function is identically equal to i, so we can assume $\nu \neq i$.

Lemma 4.2 *Let $\nu \neq i$. The transformation $z \to \gamma(z)$ takes the lower half-plane to the circle*

$$B = \{z : |z|^2 + 1 - 2\rho Imz < 0\},\ \rho = \frac{|\nu|^2 + 1}{2Im\nu}. \tag{4.6}$$

This circle has the center point $i\rho$ radius $r = \sqrt{\rho^2 - 1}$, lies entirely in the upper half-plane, contains the point $z = i$, and is invariant under the involution $z \mapsto z' = -1/z$. In addition, the points ν and $\nu' = -1/\nu$ lie on its boundary circle $L = \partial B$.

Proof By (4.2), we have

$$Im[\gamma(z)] = \frac{(1 + |z|^2)Im\nu - (1 + |\nu|^2)Imz}{|1 + \nu z|^2}.$$

Hence the image of the lower half-plane is the disc B, which lies entirely in the upper half-plane and contains the point $z = i$. By the symmetry principle, the points pmi are symmetric as relative to the straight line $\mathbb{R}$ and to the circle $L = \partial B$. In particular, the center of this circle must lie on the imaginary axis. Denoting the

center and radius of this circle, respectively, $i\rho$ and r, we come to the relation $|i - i\rho||i + i\rho| = r^2$, whence $r^2 = \rho^2 - 1$. The equation $|z - i\rho|^2 = r^2$ of the circle L can be written in the form $|z|^2 + 1 - 2\rho Imz = 0$, which proves the description (4.6) of the circle B.
Obviously, the points $\gamma(0) = \nu$ and $\gamma(\infty) = -1/\nu$ lie on L. In particular, substituting $z = \nu$, into this equation, we arrive at the expression for ρ in (4.6). The fact that the circle L is invariant under the transformation $z \mapsto z' = -1/z$ follows directly from its equation. The lemma is proved. □

Acknowledgments The research is financially supported by a grant AP 14869558.

References

1. A.V. Bitsadze, On some properties of polyharmonic functions. Differ. Equ. **24**(5), 825–831 (1988)
2. A.A. Dezin, The second boundary problem for the polyharmonic equation in the space W_2^m. Dokl. Akad. Nauk SSSR (N.S.) **96**, 901–903 (1954)
3. N.A. Malakhova, A.P. Soldatov, On a boundary value problem for a higher-order elliptic equation. Differ. Equ. **44**(8), 1111–1118 (2008). https://doi.org/10.1134/S0012266108080089
4. A.P. Soldatov, A boundary value problem for higher order elliptic equations in many connected domain on the plane. Vladikavkazskii Mat. Zh. **19**(3), 51–58 (2017)
5. B.D. Koshanov, A.P. Soldatov, Boundary value problem with normal derivatives for a higher order elliptic eguation on the plane. Differ. Equ. **52**(12), 1594–1609 (2016). https://doi.org/10.1134/S0012266116120077
6. T.Sh. Kalmenov, B.D. Koshanov, M.Y. Nemchenko, Green function representation for the Dirichlet problem of the polyharmonic equation in a sphere. Complex Variables Elliptic Equ. **53**(2), 177–183 (2008). https://doi.org/10.1080/17476930701671726
7. T.Sh. Kalmenov, B.D. Koshanov, M.Yu. Nemchenko, Green function representation in the Dirichlet problem for polyharmonic equations in a ball. Dokl. Math. **78**(1), 528–530 (2008). https://doi.org/10.1134/S1064562408040169
8. M.A. Sadybekov, B.K. Turmetov, B.T. Torebek, On an explicit form of the Green function of the Roben problem for the Laplace operator in a circle. Adv. Pure Appl. Math. **6**(3), 163–172 (2015)
9. M.A. Sadybekov, B.T. Torebek, B.K. Turmetov, Representation of Green's function of the Neumann problem for a multi-dimensional ball. Complex Var. Elliptic Equ. **61**(1), 104–123 (2016)
10. H. Begehr, T.N.H. Vu, Z.-X. Zhang, Polyharmonic Dirichlet problems. Proc. Steklov Inst. Math. **255**, 13–34 (2006)
11. H. Begehr, J. Du, Y. Wang, A Dirichlet problem for polyharmonic functions. Ann. Mat. Pura Appl. **187**(4), 435–457 (2008)
12. H. Begehr, T. Vaitekhovich, Modefied harmonic Robin function. Complex Var. Elliptic Equ. **58**(4), 483–496 (2013)
13. B.D. Koshanov, Green's functions and correct restrictions of the polyharmonic operator. Bull. KazNU. Series Math. Mech. Comput. Sci. **109**(1), 34–54 (2021). https://doi.org/10.26577/JMMCS.2021.v109.i1.03
14. A.A. Douglis, On uniqueness in Cauchy probblems for ellipilc systems of equations. Commun. Pure Appl. Math. **13**(4), 593–607 (1960)

15. S. Agmon, A. Douglis, L. Nirenberg, Estimates near the boundary for solutions of elliptic partial differential equations satisfying general boundary conditions. I. Commun. Pure Appl. Math. **12**(4), 623–727 (1959)
16. R.P. Gilbert, *Function Theoretic Methods in Partial Differential Equations* (Academic Press, New York, 1969)
17. R.Z. Yeh, Hyperholomorphic functions and higher order partial differentials equations in the plane. Pac. J. Math. **142**(2), 379–399 (1990)
18. A.P. Soldatov, Higher-order elliptic systems. Differ. Equ. **25**(1), 109–115 (1989)
19. A.P. Soldatov, On the theory of anisotropic flat elasticity. J. Math. Sci. **235**(4), 484–535 (2018)
20. A.P. Soldatov, Generalized potentials of double layer in plane theory of elasticity. Evrasian Math. J. **5**(4), 78–125 (2014)
21. S. Nazarov, B.A. Plamenevsky, *Elliptic Problems in Domains with Piecewise Smooth Boundaries* (Berlin: de Gruyter, 1994)
22. Ya.B. Lopatinskii, On a method of reducing boundary-value problems for a system of differential equations of elliptic type to regular integral equations. Ukrain. Math. J. **5**, 123–151 (1953)
23. B.D. Koshanov, A.P. Soldatov, On the solvability of the generalized Neumann problem for a high-order elliptic equation in an infinite domain. CMFD **67**(3), 564–575 (2021). http://mi.mathnet.ru/cmfd435

Solutions with Power Type or Exponential Singularity for (3+1)-D Protter-Morawetz Problems

Nedyu Popivanov, Todor Popov, and Ingo Witt

Abstract Transonic flows in fluid dynamics are modelled by mixed-type equations on the plane. On the other hand, M. H. Protter proposed multi-dimensional analogues of the classical planar Guderley-Morawetz problem for equations of hyperbolic-elliptic type. We study the Protter-Morawetz problems in the hyperbolic part of the domain. Unlike the two-dimensional analogues, for smooth right-hand sides, in the frame of classical solvability, the four-dimensional problem is not well-posed and not Fredholm. Alternatively, the uniquely determined generalized solutions may have singularities at one boundary point. We study the growth of the generalized solution at the point of singularity.

1 Introduction and History

We study some boundary value problems for the wave equations that were proposed by M. H. Protter [31, 32] in the context of models in fluid dynamics. In $\mathbb{R}^2$ the Guderley-Morawetz problem for the Gellerstedt equation of mixed hyperbolic-elliptic type is a classical model for transonic flows around airfoils (see the survey [22] by Morawetz). Morawetz [21] first established the existence of weak solutions and uniqueness of strong solutions, and a couple of years later, P. Lax and R. Phillips

N. Popivanov (✉)
Institute of Information and Communication Technologies, Bulgarian Academy of Sciences, Sofia, Bulgaria

Department of Mathematics and Informatics, University of Sofia, Sofia, Bulgaria
e-mail: nedyu@fmi.uni-sofia.bg

T. Popov
Department of Mathematics and Informatics, University of Sofia, Sofia, Bulgaria
e-mail: topopover@fmi.uni-sofia.bg

I. Witt
Mathematical Institute, University of Göttingen, Göttingen, Germany
e-mail: ingo.witt@mathematik.uni-goettingen.de

M. Ruzhansky, B. Torebek (eds.), *Extended Abstracts MWCAPDE 2023*,
Research Perspectives Ghent Analysis and PDE Center 1,
https://doi.org/10.1007/978-3-031-41665-1_20

filled a gap between weak and strong solutions. They showed in [18] that the weak solutions are strong and proved the uniqueness and existence of a strong solution to the Guderley-Morawetz problem in an appropriate weighted Sobolev space. At the same time, M.H. Protter [31] proposed some multi-dimensional analogues of the planar Guderley-Morawetz problem. Naturally, it was believed that one could adapt the methods used to study the planar case to attack the multi-dimensional problems. However, the multi-dimensional Protter-Morawetz problem proved to be qualitatively different, and the situation there is still not clear. Even the question of well-posedness is not completely resolved.

In relation to the mixed-type problem, Protter [31] studied some boundary value problems solely in the hyperbolic part of the domain. Here we consider Protter problems for the wave equation with three space variables and time

$$\Delta u - u_{tt} = f(x, t) \tag{1.1}$$

with points $(x, t) = (x_1, x_2, x_3, t)$ in the domain

$$\Omega = \{(x, t) \colon 0 < t < 1/2,\ t < |x| < 1 - t\}.$$

The domain Ω is bounded by the two characteristic cones

$$\Sigma_1 = \{(x, t) \colon 0 < t < 1/2,\ |x| = 1 - t\},$$

$$\Sigma_2 = \{(x, t) \colon 0 < t < 1/2,\ |x| = t\}$$

and the non-characteristic ball, centred at the origin $O(0, 0, 0, 0)$

$$\Sigma_0 = \{t = 0,\ |x| < 1\}.$$

In the present paper, we will study the following Protter problems $P2$ and $P2^*$.

Problem P2 Find a solution to the wave equation (1.1) in Ω which satisfies the boundary conditions

$$u_t|_{\Sigma_0} = 0, \quad u|_{\Sigma_1} = 0.$$

Its adjoint boundary value problem is as follows.

Problem P2* Find a solution to the wave equation (1.1) in Ω which satisfies the adjoint boundary conditions

$$u_t|_{\Sigma_0} = 0, \quad u|_{\Sigma_2} = 0.$$

One can think of Problems $P2$ and $P2^*$ as four-dimensional versions of the classical planar Darboux problem for the vibrating string equation in a characteristic triangle – with data prescribed on one characteristic surface and on the noncharacteristic boundary part.

Protter [32] also studied Problems $P1$ and $P1^*$ that are formulated similarly to Problems $P2$ and $P2^*$, but have Dirichlet-type boundary condition on Σ_0. Garabedian [8] proved the uniqueness of a classical solution for the four-dimensional Protter Problem $P1$. At that time, it was expected that for sufficiently smooth right-hand sides these BVPs are classically solvable. However, soon it became clear that unlike the planar Darboux problem, the Protter Problems $P1$ and $P2$ are not well posed. In fact, the homogeneous adjoint Problems $P1^*$ and $P2^*$ have an infinite number of linearly independent nontrivial classical solutions (see [17, 24, 35]).

In order to present the solutions to the homogeneous Problem $P2^*$, we will need the set of *spherical functions* Y_n^m on the unit ball $\mathbb{S}^2 := \{(x_1, x_2, x_3) : x_1^2+x_2^2+x_3^2 = 1\}$ in $\mathbb{R}^3$, with $n \in \mathbb{N} \cup \{0\}$ and $m = 1, \dots, 2n + 1$. The spherical functions form a complete orthonormal system in $L_2(\mathbb{S}^2)$, and their standard definition involves spherical coordinates (see [14]). In Cartesian coordinates, they could be given by

$$Y_n^{2p}(x) = C_n^{2p} \frac{d^p}{dx_3^p} P_n(x_3) \operatorname{Im}\left\{(x_2 + ix_1)^p\right\}, \text{ for } p = 1, \dots, n;$$

$$Y_n^{2p+1}(x) = C_n^{2p+1} \frac{d^p}{dx_3^p} P_n(x_3) \operatorname{Re}\left\{(x_2 + ix_1)^p\right\}, \text{ for } p = 0, \dots, n,$$

where P_n are the *Legendre polynomials* and C_n^m are constants. For convenience in the discussions that follow, we extend the spherical functions out of $\mathbb{S}^2$ radially, keeping the same notation Y_n^m for the extended function, i.e. $Y_n^m(x) := Y_n^m(x/|x|)$ for $x \in \mathbb{R}^3 \setminus \{O\}$.

For $n, k \in \mathbb{N} \cup \{0\}$ define for $r > 0$ and $r \geq t \geq 0$ the functions

$$E_k^n(r, t) = \sum_{i=0}^{k} B_{k,i}^n \frac{(r^2 - t^2)^{n-k-i}}{r^{n-2i+1}},$$

where the coefficients are

$$B_{k,i}^n := (-1)^i \frac{(k - i + 1)_i (n + 1 - k - i)_i}{i!(n - i + \frac{1}{2})_i}, \quad B_{k,0}^n = 1,$$

with $(a)_i := a(a + 1) \cdots (a + i - 1)$ and $(a)_0 := 1$. Then, by explicit calculations, one can show that

$$W_{k,m}^n(x, t) = \begin{cases} E_k^n(|x|, t) Y_n^m(x), & \text{in } \overline{\Omega} \setminus O; \\ 0 & \text{at } O, \end{cases}$$

are classical solutions to the homogeneous adjoint Protter problem $P2^*$.

Lemma 1.1 ([27]) *The functions $W^n_{k,m}(x,t)$ belong to $C^\infty(\overline{\Omega}\setminus O)\cap C^1(\overline{\Omega})$ and are classical solutions to the homogeneous Problem $P2^*$ for $k = 0, 1, \ldots, [n/2]-2$.*

From this result follows that the Problem $P2$ is not Fredholm solvable. Actually, the solution could be bounded only when an infinite number of orthogonality conditions for f is fulfilled – see the conditions (1.2) and Theorem 1.3. Alternatively, following [24], we introduce the notion of *generalized solutions* to the Problem $P2$ that may have singularities.

Definition 1.2 ([27]) A function $u = u(x,t)$ is called a generalized solution to Problem $P2$ in Ω if the following conditions are satisfied:

1) $u \in C^1(\overline{\Omega}\setminus\overline{\Sigma}_2)$, $u_t|_{\Sigma_0\setminus O} = 0$, $u|_{\Sigma_1} = 0$, and
2) the identity

$$\int_\Omega (u_t w_t - \nabla u\cdot\nabla w - fw)\,dxdt = 0$$

holds for all $w \in C^1(\overline{\Omega})$ such that $w_t|_{\Sigma_0} = 0$ and $w = 0$ in a neighborhood of $\overline{\Sigma}_2$.

Notice that this definition allows the generalized solution to be singular on the light cone $\overline{\Sigma}_2$. Uniqueness of the generalized solution to Problem $P2$ is proved in [27] (see also [29, Theorem 18]). According to [24], for each natural number n, there is a right-hand side $f_n \in C^n(\overline{\Omega})$, such that the corresponding generalized solution has a growth like $|x|^{-n}$ at the origin O. Let us point out that these singularities are isolated at O and do not propagate along the characteristic cone Σ_2. The difference with the classical case of propagation of singularities for hyperbolic operators (see in Hörmander [11, Chapter 24.5]) is explained by the fact that Σ_2 is part of the boundary of the domain Ω.

When the right-hand side function f is smooth, in [29, Theorem 6] the necessary and sufficient conditions for existence of bounded solutions are found.

Theorem 1.3 ([29]) *Let the function $f(x,t)$ belong to $C^6(\overline{\Omega})$. Then the necessary and sufficient conditions for existence of bounded generalized solution $u(x,t)$ to the Protter Problem $P2$ are*

$$\int_\Omega W^n_{k,m}(x,t)f(x,t)\,dxdt = 0, \tag{1.2}$$

for all $n \in \mathbb{N}\cup\{0\}$, $k = 0, \ldots, [n/2]$, $m = 1, \ldots, 2n+1$.

Moreover, this generalized solution $u(x,t) \in C^1(\overline{\Omega} \setminus O)$ *and satisfies for* $(x,t) \in \overline{\Omega} \setminus O$ *the a priori estimates*

$$|u(x,t)| \le C\left(\|f\|_{C^5(\overline{\Omega})} + \|f_t\|_{C^5(\overline{\Omega})}\right);$$

$$\sum_{i=1}^{3} \left|u_{x_i}(x,t)\right| + |u_t(x,t)| \le C(|x|^2 + t^2)^{-1} \|f\|_{C^6(\overline{\Omega})},$$

where the constant C *is independent of the function* $f(x,t)$.

We suspect that the C^6-smoothness of f is not essential, but it is needed for the technique of proof.

Consider the operator $T : u_f \longmapsto f$, where u_f is the unique classical solution to the Problem $P2$ for the right-hand side f. Obviously, the operator T is not Fredholm since $\dim \operatorname{coker}(T) = \infty$. On the other hand, the uniqueness result shows that $\dim \ker(T) = 0$. Theorem 1.3 suggests that T might be a **semi-Fredholm operator**. A semi-Fredholm operator is a bounded operator that has a finite-dimensional kernel or cokernel and closed range.

Let us point out that most authors, starting with Protter himself, discussed the Protter problems but for homogeneous wave equation with non-homogeneous boundary conditions. This makes quite difficult even the formulation of the analogue of the necessary orthogonality conditions like (1.2). We feel that it is more convenient to consider the non-homogeneous wave equations with homogeneous boundary conditions – as in Popivanov and Schneider [24].

A comparison of various recent results for Protter problems is made in [7]. For the wave equation, some other multi-dimensional variants of the planar Darboux problem are studied in [3, 4]. For Protter problems with lower-order terms or with more general boundary condition on Σ_0, see [9, 10] and the references therein. Regarding results for equations of Keldysh type, we refer to [12, 13, 25, 26, 34]. The existence of bounded or unbounded solutions for the wave equation and some other related equations is considered in [15–17, 27].

Regarding the Protter-Morawetz mixed-type hyperbolic-elliptic problems, uniqueness results for quasi-regular solutions are proved in [2]. For results concerning existence or nonexistence of nontrivial solutions to related nonlinear problems of mixed hyperbolic-elliptic type in the multi-dimensional case, see [6, 19, 20]. Uniqueness of the solution to a BVP for a hyperbolic-parabolic equation is established in [1].

2 Generalized Solutions with Power-Type Singularity

Generally, a smooth function $f(x,t)$ can be expanded as a harmonic series

$$f(x,t) = \sum_{n=0}^{\infty} \sum_{m=1}^{2n+1} f_n^m(|x|,t) Y_n^m(x)$$

of the complete orthonormal system in $L_2(\mathbb{S}^2)$ of the spherical functions Y_n^m. In [29, Theorem 4] the special case when only finite number of the Fourier coefficients f_n^m are not identically zero is considered, and for the corresponding generalized solution, the exact asymptotic formula at the point of singularity O is found. In order to formulate the result, suppose that for some number $l \in \mathbb{N} \cup \{0\}$ the right-hand side function $f \in C^1(\overline{\Omega})$ of the wave equation (1.1) has the representation

$$f(x,t) = \sum_{n=0}^{l} \sum_{m=1}^{2n+1} f_n^m(|x|,t) Y_n^m(x). \tag{2.1}$$

The asymptotic formula shows that the behaviour of the generalized solution to Problem $P2$ depend on the parameters

$$\beta_{k,m}^n := \int_{\Omega} W_{k,m}^n(x,t) f(x,t)\, dxdt,$$

where $n = 0, \ldots, l$; $k = 0, \ldots, \left[\frac{n}{2}\right]$ and $m = 1, \ldots, 2n+1$.

Theorem 2.1 ([29]) *Suppose that the right-hand side function $f \in C^1(\overline{\Omega})$ has the form (2.1). Then the unique generalized solution $u(x,t)$ to Problem $P2$ belongs to $C^2(\overline{\Omega} \setminus O)$ and has the following asymptotic expansion at the singular point O:*

$$u(x,t) = \sum_{p=1}^{l+1} \left(|x|^2 + t^2\right)^{-p/2} F_p(x,t) + F(x,t), \tag{2.2}$$

where:

(i) the function $F \in C^2(\overline{\Omega} \setminus O)$ and satisfies the a priori estimate

$$|F(x,t)| \le C \,||f||_{C^1(\Omega)}\,, \quad (x,t) \in \Omega,$$

with constant C independent of f and $||f||_{C^k(\Omega)} = \sum_{|\alpha| \le k} \max_{\overline{\Omega}} |D^\alpha f(x,t)|$;

(ii) *the functions F_p, $p = 1, \dots, l+1$, have the form*

$$F_p(x,t) = \sum_{k=0}^{[(l-p+1)/2]} \sum_{m=1}^{2p+4k-1} \beta_{k,m}^{p+2k-1} F_{k,m}^{p+2k-1}(x,t)\,, \tag{2.3}$$

with functions $F_{k,m}^n \in C^2(\overline{\Omega} \setminus O)$ bounded and independent of f;

(iii) *if at least one of the constants $\beta_{k,m}^{p+2k-1}$ in (2.3) is different from zero, then for the corresponding function $F_p(x,t)$, there exists a direction $(\alpha, 1) := (\alpha_1, \alpha_2, \alpha_3, 1)$ with $(\alpha, 1)t \in \Sigma_2$ for $0 < t < 1/2$, such that*

$$\lim_{t \to +0} F_p(\alpha t, t) = c_{p,\alpha} = const \neq 0.$$

This result shows that if only finite number of the orthogonality conditions (1.2) are not fulfilled, the solution may have at most polynomial growth, and in fact, each orthogonality condition is responsible for one power type singularity. However, in the general case of smooth right-hand side functions, one could expect that the solutions may have stronger singularities at O.

It is interesting to compare the functions $(|x|^2+t^2)^{-p/2}F_p(x,t)$ appearing in the right-hand side of (2.2) with the solutions $v_p = v_p(x,t)$ to the homogeneous wave equation (1.1) in $\Omega \cup \{t \geq |x|, 0 < t < 1/2\}$ satisfying $v_p\big|_{\Sigma_0} = |x|^{-p}F_p(x,0) \in S'(\mathbb{R}^3)$, $\partial_t v_p\big|_{\Sigma_0} = 0$. One can show that $v_p\big|_{\Sigma_0}$ is classically conormal with respect to the origin O; see [11, 33]. By a symbolic argument, one then concludes that v_p is classically conormal with respect to Σ_2. In fact, a closer inspection reveals that v_p satisfies the so-called transmission condition with respect to Σ_2 (see [5, 11, 33]), and one finds that $v_p\big|_{\{t>|x|\}}$ is C^∞ smooth up to Σ_2. In some sense v_p can be regarded as an extension of the singular part of a solution u_p to the Protter Problem $P2$ that behaves like $(|x|^2+t^2)^{-p/2}$ near the origin O. In fact, v_p will differ in Ω from u_p by a smooth function.

3 Generalized Solutions with Exponential Singularity

For smooth right-hand side functions, some estimates in [24] suggest that the unique generalized solution may have even an exponential growth. Actually, in [30] we managed to construct a smooth right-hand side, such that the corresponding solution behaves like $\exp(1/|x|)$ near the origin.

Theorem 3.1 ([30]) *There exists a right-hand side function $f(x,t) \in C^\infty(\overline{\Omega})$ and constants $C > 0$ and $\delta \in (0, 1/2)$, such that the unique generalized solution $u(x,t) \equiv u(x_1, x_2, x_3, t)$ to Problem $P2$ satisfies the estimate from above*

$$|u(x,t)| \leq C|x|^{-1}\exp\left(2(|x|+t)^{-1}\right), \quad \text{for all } (x,t) \in \Omega,$$

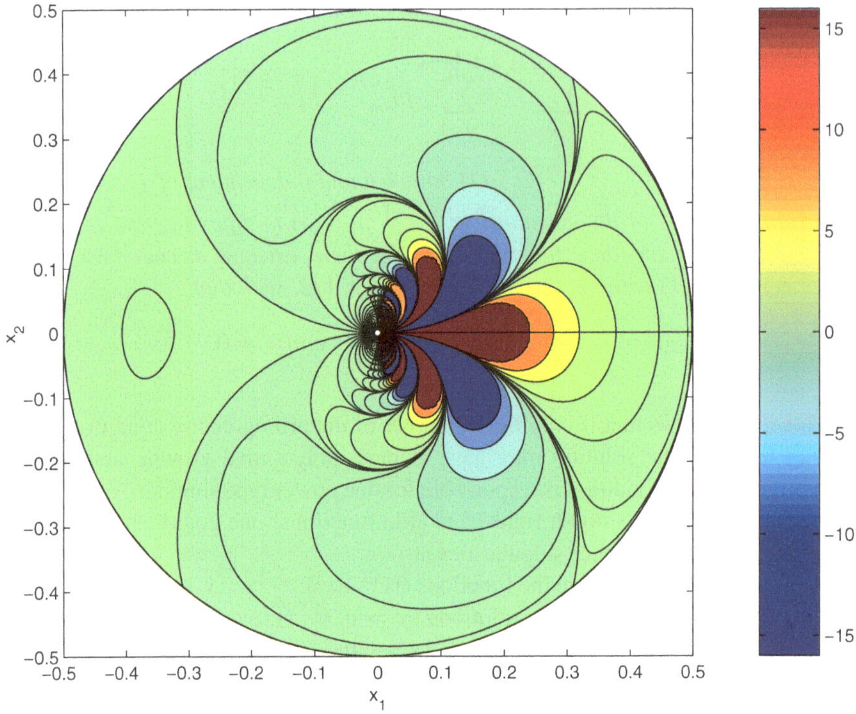

Fig. 1 A solution with exponential singularity

while on the line $\{(x_1, x_2, x_3, t) \colon x_1 = t, x_2 = x_3 = 0\}$ *near the origin* O*, the generalized solution satisfies the estimate*

$$u(t, 0, 0, t) \geq \exp\left(t^{-1}\right), \qquad \textit{for } t \in (0, \delta). \tag{3.1}$$

Figure 1 illustrates the exponential singularity on $\Sigma_2 \cap \{x_3 = 0\}$ of the generalized solution $u(x_1, x_2, x_3, t)$ constructed in Theorem 3.1. The graph shows the isolines of the function $u(x_1, x_2, 0, |x_1|)$ on the plane $(x_1, x_2) \in \mathbb{R}^2$. According to the estimate (3.1), the solution has an exponential growth on line $\{x_1 > 0, x_2 = 0\}$ (the line on the right of the graph) when approaching O (the centre of the graph). It looks that the solution changes its sign infinite times near O in $\{x_1 > 0\}$. This raises the question if this solution has an even faster growth than shown in (3.1) in some directions.

Finally, we will formulate some open questions that naturally arise from the results discussed above.

Open problems

It is interesting to find out why such a strong singularity appears at a single isolated point, which is not typical of the wave equation. What is the reason for this situation mathematically? Are there some physical explanations for this situation? An essential task is to study such phenomena when the solution is smooth in $\overline{\Omega} \setminus O$ and has an exponential singularity only at the point O.

In the frame of the distributions theory, it is impossible to extend the function $u = u(x, t)$ constructed in Theorem 3.1 as a generalized solution to the wave equation across the light cone Σ_2. We plan to discuss this and related questions in a future publication.

Acknowledgments The work of N. Popivanov and T. Popov was partially supported by Bulgarian NSF under Grant KP-06-R-126/2020.

References

1. S.A. Aldashev, Uniqueness of the solution of the Tricomi-type problem for multidimensional hyperbolic-parabolic equation. Differ. Equ. **53**(1), 133–138 (2017)
2. A.K. Aziz, M. Schneider, Frankl-Morawetz problems in R^3. SIAM J. Math. Anal. **10**, 913–921 (1979)
3. Ar.B. Bazarbekov, Ak.B. Bazarbekov, The Goursat and Darboux problems for the three-dimensional wave equation. Differ. Equ. **38**, 695–701 (2002)
4. A.V. Bitsadze, *Some Classes of Partial Differential Equations* (Gordon and Breach Science Publishers, New York, 1988)
5. L. Boutet de Monvel, Lacunas and transmissions, in *Seminar on Singularities of Solutions of Linear Partial Differential Equations*, ed. by L. Hörmander. Annuals of Mathematics Studies, vol. 91 (Princeton University Press, Princeton, NJ, 1978), pp. 209–218
6. L. Dechevski, K.R. Payne, N. Popivanov, Nonexistence of nontrivial generalized solutions for 2-D and 3-D BVPs with nonlinear mixed type equations. AIP Conf. Proc. **1910**(040015), 1–13 (2017)
7. L. Dechevsky, N. Popivanov, T. Popov, Exact asymptotic expansion of singular solutions for (2+1)-D Protter problem. Abstr. Appl. Anal. **2012**, 1–33 (2012). Article ID 278542
8. P.R. Garabedian, Partial differential equations with more than two variables in the complex domain. J. Math. Mech. **9**, 241–271 (1960)
9. M.K. Grammatikopoulos, T.D. Hristov, N.I. Popivanov, Singular solutions to Protter's problem for the 3-D wave equation involving lower order terms. Electron. J. Diff. Eqns. **2003**(03), 1–31 (2003)
10. M.K. Grammatikopoulos, N.I. Popivanov, T.P. Popov, New singular solutions of Protter's problem for the $3 - D$ wave equation. Abstr. Appl. Anal. **2004**(4), 315–335 (2004)
11. L. Hörmander, *The Analysis of Linear Partial Differential Operators III: Pseudo-Differential Operators*. Grundlehren der mathematischen Wissenschaften, vol. 274 (Springer, Berlin, 1985)
12. T. Hristov, Singular solutions to Protter problem for Keldysh type equations. AIP Conf. Proc. **1631**, 255–262 (2014)
13. T. Hristov, A. Nikolov, Behaviour of singular solutions to Protter problem for (3+1)-D Keldysh type equations. Compt. Rend. Acad. Bulg. Sci. **70**(2), 167–174 (2017)
14. M.N. Jones, *Spherical Harmonics and Tensors for Classical Field Theory* (Research Studies Press, Letchworth, 1986)
15. J.D. Jeon, K.K. Cher, J.H. Park, Y.H. Jeon, J.B. Choi, Protter's conjugate boundary value problems for the two dimensional wave equation. J. Korean. Math. Soc. **33**, 857–863 (1996)
16. J.B. Choi, J.Y. Park, On the conjugate Darboux-Protter problems for the two dimensional wave equations in the special case. J. Korean Math. Soc. **39**(5), 681–692 (2002)

17. K.K. Cher, Nontrivial solutions of some homogeneous boundary value problems for a many-dimentional hyperbolic Euler-Poisson-Darboux equation in an unbounded domain. Differ. Equ. **34**(1), 139–142 (1998)
18. P.D. Lax, R.S. Phillips, Local boundary conditions for dissipative symmetric linear differential operators. Commun. Pure Appl. Math. **13**, 427–455 (1960)
19. D. Lupo, K. Payne, N. Popivanov, Nonexistence of nontrivial solutions for supercritical equations of mixed elliptic-hyperbolic type, in *Contributions to Nonlinear Analysis, Progress in Non-Linear Differential Equations and Their Applications* (Birkhauser, Basel, 2006), pp. 371–390
20. D. Lupo, K. Payne, N. Popivanov, On the degenerate hyperbolic Goursat problem for linear and nonlinear equations of Tricomi type. Nonlinear Anal.: Theory Methods Appl. **108**, 29–56 (2014)
21. C.S. Morawetz, A weak solution for a system of equations of elliptic-hyperbolic type. Commun. Pure Appl. Math. **11**, 315–331 (1958)
22. C.S. Morawetz, Mixed equations and transonic flow. J. Hyperbolic Differ. Equ. **1**(1), 1–26 (2004)
23. A. Nikolov, Integral form of the generalized solution of a Darboux-Goursat problem with third-type boundary condition. AIP Conf. Proc. **2333**, 120009 (2021)
24. N. Popivanov, M. Schneider, On M. H. Protter problems for the wave equation in $\mathbb{R}^3$. J. Math. Anal. Appl. **194**, 50–77 (1995)
25. N. Popivanov, T. Hristov, A. Nikolov, M. Schneider, On the existence and uniqueness of a generalized solution of the Protter problem for (3+1)-D Keldysh-type equations. Bound. Value Prob. **2017**, 26:1–30 (2017)
26. N. Popivanov, T. Hristov, A. Nikolov, M. Schneider, Singular solutions to a (3+1)-D Protter-Morawetz problem for Keldysh-type equations. Adv. Math. Phys. **2017**, 1–16 (2017). Article ID 1571959
27. N. Popivanov, T. Popov, Singular solutions of Protter's problem for the $3+1$-D wave equation. Integral Transform. Spec. Funct. **15**(1), 73–91 (2004)
28. N. Popivanov, T. Popov, R. Scherer, Protter-Morawetz multidimensional problems. Proc. Steklov Inst. Math. **278**, 179–198 (2012)
29. N. Popivanov, T. Popov, A. Tesdall, Semi-Fredholm solvability in the framework of singular solutions for the (3+1)-D Protter-Morawetz problem. Abstr. Appl. Anal. **2014**, 1–19 (2014). Article ID 260287
30. N. Popivanov, T. Popov, I. Witt, Solutions with exponential singularity for (3+1)-D Protter problems, J. Hyperbolic Differ. Equ. **20**(02), 475–498 (2023)
31. M.H. Protter, New boundary value problems for the wave equation and equations of mixed type. J. Rat. Mech. Anal. **3**, 435–446 (1954)
32. M.H. Protter, A boundary value problem for the wave equation and mean value problems. Ann. Math. Stud. **33**, 247–257 (1954)
33. Z.-P. Ruan, I. Witt, A scheme for solving hyperbolic problems with symbolic structure, to appear in *Differential Integral Equations*
34. C. Shuxing, A mixed equation of Tricomi-Keldysh type. J. Hyperbolic Differ. Equ. **9**(3), 545–553 (2012)
35. K.-C. Tong, On a boundary-value problem for the wave equation. Sci. Rec. N. Ser. **1**(1), 1–3 (1957)

Boundary Value Problems for Differential Equations with a High-Order Derivative in the Boundary Condition

Makhmud Sadybekov and Batirkhan Turmetov

Abstract The paper considers boundary-value problems for the Laplace equation, the boundary conditions of which contain traces of high-order derivatives on the boundary of the domain. It is shown that there are classes of (singular) solutions in which the solution of problems exists for any (sufficiently smooth) input data, without applying conditions such as orthogonality conditions. An example of a well-posed problem, the solution of which exists and is unique, is given.

1 Introduction

Let Ω be a simple connected domain from R^n with the smooth boundary $\partial\Omega$, and $A(x, D)$ be a uniformly elliptic operator of order $2m$. Consider in Ω the following boundary value problem:

$$A(x, D)u(x) = 0, \ x \in \Omega, \tag{1.1}$$

$$B_j(x, D)u(x) = \varphi_j(x), \ x \in \partial\Omega. \tag{1.2}$$

If the operators $B_j(x, D)$ have different orders and do not exceed $2m - 1$, then problem (1.1)–(1.2) is solvable under the Lopatinskii condition, and the kernel and cokernel of the problem are finite-dimensional [1]. If the Lopatinskii condition is violated, additional conditions for the solvability of the problem arise. Such

M. Sadybekov (✉)
Institute of Mathematics and Mathematical Modeling, Almaty, Kazakhstan
e-mail: sadybekov@math.kz

B. Turmetov
Akhmet Yasawi International Kazakh-Turkish University, Turkistan, Kazakhstan
e-mail: batirkhan.turmetov@ayu.edu.kz

M. Ruzhansky, B. Torebek (eds.), *Extended Abstracts MWCAPDE 2023*,
Research Perspectives Ghent Analysis and PDE Center 1,
https://doi.org/10.1007/978-3-031-41665-1_21

problems include degenerate boundary value problems with an oblique derivative [2]:

$$\Delta u(x, y) = 0, \ x^2 + y^2 < 1,$$

$$u_x(x, y) = f(x, y), \ x^2 + y^2 = 1.$$

When some operators $B_j(x, D)$ have an order greater than $2m-1$, then additional orthogonality conditions arise for boundary functions.

In the works of Bitsadze [3] and Sokolovskij [4], in the ball $\Omega = \{r < 1\}$ the following problem was considered for harmonic functions:

$$\Delta u(x) = 0, \ r < 1, \tag{1.3}$$

$$\frac{\partial^k u(x)}{\partial r^k} = \varphi(x), \ r = 1, \tag{1.4}$$

where $k \geq 2$ and $r = \|x\|$ is a normal vector. It was shown that boundary value problem (1.3)–(1.4) *is Fredholm*. A necessary and sufficient condition of its solvability is the orthogonality of the boundary function to homogeneous harmonic polynomials of order $k - 1$.

Further, these results were generalized in the case of half-spaces for the boundary operators $B_j(x, D)$ of integer order in the work of Karachik [5] and of partial order α in the works of Umarov [6–8]. In [7] it is shown that for the solvability of the problem, in the Sobolev space starting with the order $\alpha > 1/2$, an additional orthogonality condition is necessary.

A similar result, when the domain $\Omega = \{r < 1\}$ is a ball for a polyharmonic equation with boundary operators depending on polynomials of order m_j, was obtained in the work of Karachik [9]. A detailed description of such problems can be found in the monograph of Karachik [10].

It should be noted that, apparently, Tikhonov was first who began to investigate such problems [11]. Many other works have been devoted to elliptic problems with setting a high derivative on the boundary [12–18] and others.

In the works of Turmetov, the problem

$$\Delta u(r, \varphi) = 0, \ r < 1,$$

$$B_\alpha[u](r, \varphi)\big|_{r=1} = f(\varphi), \ \ 0 \leq \varphi \leq \pi,$$

with the fractional differentiation operator B_α (in the sense of Riemann-Liouville [19] or Caputo [20]) acting in the direction of the vector r, was considered in the unit disk $\Omega = \{r < 1\}$. The order of the fractional derivative was represented as $\alpha = n + \mu, \ 0 < \mu < 1, \ n = 0, 1, \ldots$. A complete investigation was given of what happens when α changes from 0 to ∞ indicating exact orthogonality conditions and the order of smoothness of solutions in classes of Hölder C^λ [19] and of Nikiolskii H_p^s, $1 \le p \le \infty$, $s > 0$ [20] depending on values of α.

2 How Do Orthogonality Conditions Arise?

In the unit disk $\Omega = \{r < 1\}$, consider a problem for the harmonic functions

$$\Delta u(r, \varphi) = 0, \ r < 1, \tag{2.1}$$

$$\frac{\partial^2 u(r, \varphi)}{\partial r^2} = f(\varphi), \ r = 1, \ 0 \le \varphi \le 2\pi. \tag{2.2}$$

Denoting

$$v(r, \varphi) = r\frac{\partial}{\partial r}\left(r\frac{\partial}{\partial r} - 1\right)u(r, \varphi) \equiv r^2\frac{\partial^2}{\partial r^2}u(r, \varphi),$$

we get the Dirichlet problem

$$\Delta v(r, \varphi) = 0, \ r < 1, \tag{2.3}$$

$$v(r, \varphi) = f(\varphi), \ r = 1, \ 0 \le \varphi \le 2\pi, \tag{2.4}$$

and additional necessary conditions

$$v(0) = 0, \ \frac{\partial v}{\partial x_1}(0) = 0, \ \frac{\partial v}{\partial x_2}(0) = 0. \tag{2.5}$$

The solution of the Dirichlet problem (2.3)–(2.4) exists for all smooth enough $f(\varphi)$, is unique, and can be written in the form

$$v(x) = \frac{1}{\pi}\int_{\partial\Omega}\frac{1 - |x|^2}{|x - y|^2}f(y)dS_y. \tag{2.6}$$

Substituting this into the necessary additional conditions (2.5), we have

$$v(0) = 0 \Longleftrightarrow \int_{\partial\Omega} f(y)dS_y = 0,$$

$$\frac{\partial v}{\partial x_j}(0) = 0 \Longleftrightarrow \int_{\partial\Omega} y_j f(y)dS_y = 0.$$

In polar coordinates, these conditions are written as

$$\int_0^{2\pi} f(\theta)d\theta = 0, \quad \int_0^{2\pi} \begin{Bmatrix} \cos(\theta) \\ \sin(\theta) \end{Bmatrix} f(\theta)d\theta = 0. \tag{2.7}$$

Thus, the solution of problem (2.1)–(2.2) exists under the orthogonality conditions (2.7). And this solution has the form

$$u(r,\varphi) = \int_0^1 (1-s)s^{-2} v(sr,\varphi)ds + C_0 + C_1 r\cos\varphi + C_2 r\sin\varphi, \tag{2.8}$$

where the function $v(r,\varphi)$ is represented in the form

$$v(r,\varphi) = \frac{1}{\pi}\int_{-\pi}^{\pi} \frac{1-r^2}{1-2r\cos(\varphi-\theta)+r^2} f(\theta)d\theta.$$

3 Orthogonality Conditions by the Method of Separation of Variables

As is well-known, if the solution is sought in the form of a trigonometric series

$$u(r,\varphi) = R_0(r)a_0 + \sum_{n=1}^{\infty} R_n(r)\Big\{a_n\cos(n\varphi) + b_n\sin(n\varphi)\Big\},$$

then for the unknown functions $R_n(r)$, we obtain equations

$$r^2 R_n''(r) + rR_n'(r) = n^2 R_n(r),$$

whose general solutions are

$$R_0(r) = A_0 + B_0\ln r, \quad R_n(r) = A_1 r^n + B_1 r^{-n}.$$

To obtain the solution which is "good" at the point 0, we assume $B_0 = 0$ and $B_1 = 0$.

As a result, the solution has the form

$$u(r,\varphi) = a_0 + \sum_{n=1}^{\infty} r^n\Big\{a_n\cos(n\varphi) + b_n\sin(n\varphi)\Big\}.$$

Let us differentiate the solution

$$u(r,\varphi) = a_0 + \sum_{n=1}^{\infty} r^n \Big\{a_n \cos(n\varphi) + b_n \sin(n\varphi)\Big\}$$

twice with respect to r

$$\frac{\partial^2 u(r,\varphi)}{\partial r^2} = \sum_{n=2}^{\infty} n(n-1)r^{n-2}\Big\{a_n \cos(n\varphi) + b_n \sin(n\varphi)\Big\}.$$

Tending $r \to 1$, from the boundary condition (2.2), we obtain

$$f(\varphi) = \sum_{n=2}^{\infty} n(n-1)\Big\{a_n \cos(n\varphi) + b_n \sin(n\varphi)\Big\}.$$

Hence we see the necessity of the orthogonality conditions (2.7).

4 Problem in the Class of Singular Solutions

Thus, from the results of Sects. 2 and 3, it follows that there is no solution to problems of this type:

$$\Delta\widetilde{u}(r,\varphi) = 0, \ r < 1, \tag{4.1}$$

$$\left.\frac{\partial^2 \widetilde{u}(r,\varphi)}{\partial r^2}\right|_{r=1} = c_0 + \alpha_1 \cos\varphi + \beta_1 \sin\varphi, \ \ 0 \le \varphi \le 2\pi. \tag{4.2}$$

However, if we consider this problem in the class of functions that admit singularities at the point 0, then its solution exists:

$$\widetilde{u}(r,\varphi) = -c_0 \ln r + \frac{1}{2} r^{-1}\Big\{\alpha_1 \cos\varphi + \beta_1 \sin\varphi\Big\}. \tag{4.3}$$

Indeed, it is easy to verify that we have

$$\frac{\partial^2 \widetilde{u}(r,\varphi)}{\partial r^2} = c_0 r^{-2} + r^{-3}\Big\{\alpha_1 \cos\varphi + \beta_1 \sin\varphi\Big\},$$

and the fulfillment of the boundary condition (4.2) is obvious.

Note that the constructed solution (4.3) of problem (4.1)–(4.2) has a singularity at the point 0 of the form $\ln r$ and r^{-1}. The order of this singularity is higher than the singularity of the fundamental solution of the Laplace operator. Therefore, such a singularity is not a removable singularity. We call the class of solutions with such a singularity *the class of singular solutions*.

It should be noted that there are several works known to us on the study of boundary value problems for the Laplace equation in a "punctured" domain. A domain is called "*punctured*" if it contains points, where the differential equation does not hold. Moreover, the singularity of the solution at these points is not the removable singularity. Among such works, we highlight the works of Kanguzhin [21–25] as the closest to the problem we are considering.

5 Conditions for the Uniqueness of the Solution of the Problem

As can be seen from the representation of the solution in the form (2.8), the solution of problem (2.1)–(2.2) is not unique. It is determined up to the component

$$a_0 + a_1 r \cos\varphi + b_1 r \sin\varphi. \tag{5.1}$$

In other words, the corresponding homogeneous problem has a three-parameter family of solutions of form (5.1).

Therefore, for the uniqueness of the solution, it suffices to specify the value of the solution at three (correctly chosen) points $A_j \in \overline{\Omega}$, $j = 0, 1, 2$.

For example, we can specify at the points $A_j = \left(r_0, j\frac{\pi}{2}\right)$, $0 < r_0 \leq 1$:

$$u\left(r_0, j\frac{\pi}{2}\right) = f_j, \ j = 0, 1, 2.$$

Then, to determine the unknown coefficients, we obtain the system

$$\begin{pmatrix} 1 & r_0 & 0 \\ 1 & 0 & r_0 \\ 1 & -r_0 & 0 \end{pmatrix} \begin{pmatrix} a_0 \\ a_1 \\ b_1 \end{pmatrix} = \begin{pmatrix} f_0 \\ f_1 \\ f_2 \end{pmatrix}, \ \Delta = 2r_0^2 \neq 0,$$

whose solution exists and is unique.

It is obvious that we can indicate many such problem statements.

6 On One Well-Posed Problem with a High Derivative on the Boundary of the Domain

Let us formulate one boundary value problem that has a high-order derivative on the boundary and is well-posed in the class of singular solutions.

Problem A Find a function $u(r, \varphi)$ that is harmonic in the disc $|r| < 1$ except for the point 0 and has a singularity of order at most $\frac{1}{r}$ at the point 0 and at the boundary satisfies the following boundary condition:

$$\frac{\partial^2 u(r, \varphi)}{\partial r^2} = f(\varphi), \ r = 1, \ 0 \leq \varphi \leq 2\pi,$$

and condition at three points

$$u\left(r_0, j\frac{\pi}{2}\right) = f_j, \ j = 0, 1, 2.$$

Let us formulate the main result on the well-posedness of this problem.

Theorem 6.1 *For any function $f \in C^\lambda(\partial\Omega)$ $(0 < \lambda < 1)$ and any set of constants f_0, f_1, f_2, there is a unique solution of Problem A. This solution belongs to the class $C^{2+\lambda}(\overline{\Omega} \setminus \{0\})$. And this solution can have a singularity at the point 0 of the form $\ln r$ and r^{-1}.*

Note that this technique can be easily transferred to boundary value problems (1.3)–(1.4) with a high derivative on the boundary and considered in a multidimensional domain from R^d.

Acknowledgments The research is financially supported by the Science Committee of the Ministry of Science and Higher Education of the Republic of Kazakhstan (Grant No AP09260752).

References

1. Y.B. Lopatinskii, On one method for the reduction of boundary-value problems for systems of elliptic differential equations to regular integral equations. Ukr. Mat. Zh. **5**(2), 123–151 (1953)
2. S.Z. Levendorskij, B.P. Paneyakh, Degenerate elliptic equations and boundary problems (Russian). Itogi Nauki Tekh., Ser. Sovrem. Probl. Mat., Fundam. Napravleniya **63**, 131–200 (1990)
3. A.V. Bitsadze, On the Neumann problem for harmonic functions. Dokl. Math. **41**(2), 193–195 (1990)
4. V.B. Sokolovskij, On a generalization of the Neumann problem (Russian). Differ. Uravn. **24**(4), 714–716 (1988)
5. V.V. Karachik, A generalized Neumann problem for harmonic functions in the half-space. Differ. Equ. **35**(7), 949–955 (1999)
6. S.R. Umarov, Critical values of the order of boundary operators in boundary-value problems for elliptic equations. Russ. Math. Surv. **46**(6), 212 (1991)
7. S.R. Umarov, On fractional derivatives of harmonic functions with a given trace (Russian). Reports of the Academy of Sciences of the Uzbek SSR, No. 10–11 (1992), pp. 17–19
8. S.R. Umarov, Some boundary value problems for elliptic equations with a boundary operator of fractional order. Dokl. Math. **48**(3), 655–658 (1994)
9. V.V. Karachik, A problem for the polyharmonic equation in the sphere. Sib. Math. J. **32**(5), 767–774 (1991)

10. V.V. Karachik, Method of normalized systems of functions (Russian) Chelyabinsk: Izd. Tsentr Yuzhno-Ural'skiĭ Gosudarstvennyĭ Universitet (2014), p. 452
11. A.N. Tikhonov, On boundary conditions containing derivatives of order greater than the order of the equation. Mat. Sb. **26**, 35–56 (1950)
12. I.I. Bavrin, Operators for harmonic functions and applications. Differ. Equ. **21**, 9–15 (1985)
13. D. Amanov, On a generalization of the Dirichlet problem for the Poisson equation. Bound. Value Probl. **2016**, 160 (2016)
14. D. Amanov, G. Ibragimov, A. Kılıçman, On a generalization of the initial-boundary problem for the vibrating string equation. Symmetry **11**(1), 73 (2019)
15. A. Kılıçman, A. Ömer, On higher-order boundary value problems by using differential transformation method with convolution terms. J. Frankl. Inst. **351**, 631–642 (2014)
16. C.H.C. Hussin, A. Kılıçman, On the solutions of non-linear higher-order boundary value problems by using differential transformation method and Adomian decomposition method. Math. Probl. Eng. **2011**, 724927 (2011)
17. R. Ashurov, Y. Fayziev, On some boundary value problems for equations with boundary operators of fractional order. Int. J. Appl. Math. **34**(2), 283–295 (2021)
18. M.A. Sadybekov, B.Kh. Turmetov, B.T. Torebek, On an explicit form of the Green function of the Robin problem for the Laplace operator in a circle. Adv. Pure Appl. Math. **6**(3), 163–172 (2015)
19. B.Kh. Turmetov, On a boundary value problem for the harmonic equation. Differ. Equ. **32**(8), 1093–1096 (1996)
20. B.K. Turmetov, On smoothness of a solution to a boundary value problem with fractional order boundary operator. Sib. Adv. Math. **15**(2), 115–125 (2005)
21. B.E. Kanguzhin, K.S. Tulenov, Correctness of the definition of the Laplace operator with delta-like potentials. Complex Var. Elliptic Equ. **67**(4), 898–920 (2022)
22. B.E. Kanguzhin, N.E. Tokmagambetov, Resolvents of well-posed problems for finite-rank perturbations of the polyharmonic operator in a punctured domain. Sib. Math. J. **57**, 265–273 (2016)
23. B.E. Kanguzhin, D.B. Nurakhmetov, N.E. Tokmagambetov, Laplace operator with δ-like potentials. Russian Math. (Iz. VUZ) **58**(2), 6–12 (2014)
24. B.E. Kanguzhin, A.A. Anijarov, Well-posed problems for the laplace operator in a punctured disk. Math. Notes **89**(6), 819–829 (2011)
25. G.E. Berikhanova, B.E. Kanguzhin, Resolvents of finite-dimensional perturbations of the correct problems for the biharmonic operator. Ufim. Mat. Zh. **2**(1), 17–34 (2010)

A Cauchy Problem for q-Heat Equation

Serikbol Shaimardan

Abstract In this paper, we explore the weak solutions of the Cauchy problem and an inverse source problem for the heat equation in the quantum calculus, formulated in abstract Hilbert spaces. For this, we use the Fourier series expansions. Moreover, we prove the existence, uniqueness, and stability of the weak solution of this problem. For a more detailed analysis and for the proofs of the announced results, we refer to (Ruzhansky M, Shaimardan S, Direct and inverse source problems for heat equation in quantum calculus. https://doi.org/10.48550/arXiv.2212.07403).

1 Introduction

Nowadays, the theory of quantum groups and q-deformed algebras have been the subject of intense investigation. Many physical applications have been investigated on the basis of the q-deformation of the Heisenberg algebra [1] and [2]. For instance, the q-deformed Schrödinger equations have been proposed in [3] and [4], and applications to the study of the q-deformed version of the hydrogen atom and of the quantum harmonic oscillator [5] have been introduced. The origin of the q-difference calculus can be traced back to the works [6, 7] by F. Jackson and R. D. Carmichael [8] from the beginning of the twentieth century, while basic definitions and properties can be found, e.g., in the monographs [9] and [10].

The paper is organized as follows: The main results are presented and proved in Sect. 3. In the final section, examples are given. In order to simplify these presentations, we include in Sect. 2 the necessary preliminaries. For a more detailed analysis and for the proofs of the announced results, we refer to [11].

S. Shaimardan (✉)
L. N. Gumilyov Eurasian National University, Astana, Kazakhstan

Department of Mathematics: Analysis, Logic and Discrete Mathematics, Ghent University, Ghent, Belgium

M. Ruzhansky, B. Torebek (eds.), *Extended Abstracts MWCAPDE 2023*,
Research Perspectives Ghent Analysis and PDE Center 1,
https://doi.org/10.1007/978-3-031-41665-1_22

2 Preliminaries

In this section, we recall some notations related to the q-calculus. We will always assume that $0 < q < 1$. The q-real number $[\alpha]_q$ is defined by $[\alpha]_q := \frac{1-q^\alpha}{1-q}$. The q-analogue differential operator $D_q f(x)$ is defined by

$$D_q f(x) = \frac{f(x) - f(qx)}{x(1-q)}. \tag{2.1}$$

The q-derivative of a product of two functions has the form

$$D_q(fg)(x) = f(qx)D_q(g)(x) + D_q(f)(x)g(x). \tag{2.2}$$

The q-integral (or Jackson integral) is defined by (see [7])

$$\int_0^x f(t)d_qt = (1-q)x\sum_{m=0}^{\infty} q^m f(xq^m). \tag{2.3}$$

Note that

$$\int_a^x D_q f(t)d_qt = f(x) - f(a). \tag{2.4}$$

The q-version of the integration by parts takes the form

$$\int_a^b f(x)D_q g(x)d_qx = [fg]_a^b - \int_a^b g(qx)D_q f(x)d_qx. \tag{2.5}$$

Let H be a separable Hilbert space and let $\mathcal{L}$ be an operator with the discrete spectrum and eigenfunctions $\{\phi_k\}_{k\in I}$ in H, where I is a countable set ($I = \mathbb{N}^k$ or $I = \mathbb{Z}^k$ for some k). We assume that the operator $\mathcal{L}$ is diagonalizable (can be written in the infinite dimensional matrix form) with respect to the orthonormal basis $\{\phi_k\}_{k\in I}$ of H with eigenvalues λ_k such that $\lambda_k > \lambda_0 > 0$, where λ_0 is some constant independent of k. In further calculus for our analysis, we will also require that $\lambda_k \to \infty$. The Plancherel identity takes the form

$$\|u\|_H = \left(\sum_{k\in I} |\langle u, \phi_k\rangle_H|^2\right)^{\frac{1}{2}}, \tag{2.6}$$

for $u \in H$.

We denote the domain of the operator $\mathcal{L}^m$ for $m \in \mathbb{N}_0$ by

$$Dom(\mathcal{L}^m) := \left\{u \in H : \mathcal{L}^i u \in Dom(\mathcal{L}), \quad i = 0, 1, 2, \cdots, m-1\right\}.$$

The space $C_{\mathcal{L}}^{\infty}(H) := \bigcap_{m=1}^{\infty} Dom(\mathcal{L}^m)$ is called the space of test functions for $\mathcal{L}$. Moreover, we introduce the Fréchet topology of $C_{\mathcal{L}}^{\infty}(H)$ with the family of norms $\|u\|_{C_{\mathcal{L}}^m} := \max_{j \leq k} \|\mathcal{L}^j u\|_H$, for $m \in \mathbb{N}_0$, $u \in C_{\mathcal{L}}^{\infty}(H)$, and the space of $\mathcal{L}$-distributions $\mathcal{D}'_{\mathcal{L}}(H) := \mathfrak{L}\left(C_{\mathcal{L}}^{\infty}(H), \mathbb{C}\right)$. Consequently, we can also define the Sobolev spaces $\mathcal{H}_{\mathcal{L}}^d$ associated with $\mathcal{L}$ as $\mathcal{H}_{\mathcal{L}}^d := \left\{u \in \mathcal{D}'_{\mathcal{L}}(H) : \mathcal{L}^{d/2} u \in H\right\}$ for any $d \in \mathbb{R}$. Using Plancherel's identity (2.6), we can write the norm in the following form:

$$\|u\|_{\mathcal{H}_{\mathcal{L}}^d} := \|\mathcal{L}^{d/2} u\|_H = \left(\sum_{k \in I} \lambda_k^d \, |\langle u, \phi_k \rangle_H|^2\right)^{\frac{1}{2}}. \tag{2.7}$$

We introduce the spaces $C_q^m\left([0, T]; \mathcal{H}_{\mathcal{L}}^d\right)$ and $L^{\infty}\left([0, T]; \mathcal{H}_{\mathcal{L}}^d\right)$ as

$$\|u\|_{C_q^m([0,T];\mathcal{H}_{\mathcal{L}}^d)} := \sum_{i=0}^{m} \max_{0 \leq t \leq T} \|D_q^i u(t)\|_{\mathcal{H}_{\mathcal{L}}^d}$$

and $\|u\|_{L^{\infty}([0,T];\mathcal{H}_{\mathcal{L}}^d)} := \sup_{0 \leq t \leq T} \|u(t)\|_{\mathcal{H}_{\mathcal{L}}^d}$, respectively, where $D_q^k u = D_q\left(D_q^{k-1} u\right)$. As given in [9], two q-analogues of the exponential function are defined as

$$e_q^x = \sum_{k=0}^{\infty} \frac{x^k}{[k]_q}, \qquad E_q^x = \sum_{k=0}^{\infty} q^{k(k-1)/2} \frac{x^k}{[k]_q}. \tag{2.8}$$

We can rewrite $E_q^x = \lim_{N \to \infty} \prod_{i=0}^{N} \left(1 + (1-q) q^i x\right)$, and we have $e_q^x E_q^{-x} = 1$.

We will be making the following assumption.

Assumption 2.1 *We assume that υ is a continuous function on $[0, T]$ and there are some $\beta, \alpha > 0$ such that $\alpha \leq \upsilon(t) \leq \beta$ for all $t \in [0, T]$.*

For $0 \leq t \leq T$ we denote

$$\gamma_{\upsilon}(t) := \frac{1}{\lim_{N \to \infty} \prod_{i=0}^{N} \left(1 + (1-q) t q^i \upsilon(q^i t)\right)} \tag{2.9}$$

and it can be written as

$$\frac{1}{E_q^{\beta t}} \leq \gamma_{\upsilon}(t) \leq \frac{1}{E_q^{\alpha t}}. \tag{2.10}$$

From (2.1) and (2.4). it follows that

$$D_q\left[\gamma_{\upsilon}^{-1}(\lambda_k t)\right] = \lambda_k \upsilon(t)\gamma_{\upsilon}^{-1}(q\lambda_k t) \tag{2.11}$$

and

$$\int_0^T \gamma_{\upsilon}^{-1}(\lambda_k s)d_q s = \frac{\gamma_{\upsilon}^{-1}(\lambda_k T)}{\lambda_k \beta} \tag{2.12}$$

and

$$\int_0^t \gamma_{\upsilon}^{-1}(\lambda_k s)d_q s = \frac{\gamma_{\upsilon}^{-1}(\lambda_k t)}{\lambda_k \alpha}. \tag{2.13}$$

Notation The symbol $M \lesssim K$ means that there exists $\gamma > 0$ such that $M \leq \gamma K$, where γ is a constant.

3 A Cauchy for the q-Heat Equation

We start to study a Cauchy problem

$$D_{q,t}u(t) + \upsilon(t)\mathcal{L}u(t) = f(t), \quad u \in H, \quad t > 0, \tag{3.1}$$

with the initial condition

$$u(0) = \varphi \in H. \tag{3.2}$$

Theorem 3.1 *Assume that Assumption 2.1 holds. Let $d \in \mathbb{R}$, $\varphi \in \mathcal{H}_{\mathcal{L}}^{d+2}$ and $f \in L^{\infty}\left([0, T]; \mathcal{H}_{\mathcal{L}}^{d+2}\right)$. Then there exists a unique solution of Problem (3.1)–(3.2):*

$$u \in C_q^1\left([0, T]; \mathcal{H}_{\mathcal{L}}^{d+2}\right) \cap L^{\infty}\left([0, T]; \mathcal{H}_{\mathcal{L}}^{d+2}\right). \tag{3.3}$$

Moreover, this solution can be written in the form

$$u(t) = \sum_{k \in I} \Bigg[\langle \varphi, \phi_k \rangle_H \gamma_\upsilon(\lambda_k t) + \gamma_\upsilon(\lambda_k t) \int_0^t \gamma_\upsilon^{-1}(q\lambda_k s) \langle f(s), \phi_k \rangle_H d_q s \Bigg] \phi_k, \tag{3.4}$$

which satisfies the estimate

$$\|u(t)\|_{\mathcal{H}_{\mathcal{L}}^{d+2}} + \|D_q u(t)\|^2_{\mathcal{H}_{\mathcal{L}}^{d}} \lesssim \|\varphi\|^2_{\mathcal{H}_{\mathcal{L}}^{d+2}} + \|f(t)\|^2_{\mathcal{H}_{\mathcal{L}}^{d+2}}. \tag{3.5}$$

Proof *Existence.* We start to solve equation (3.1). We can use the system of eigenfunctions and look for a solution in the series form

$$u(t) = \sum_{k \in I} u_k(t) \phi_k, \tag{3.6}$$

for each fixed $t > 0$. Such an expansion always exists due to the completeness of the set of eigenfunctions $\{\phi_k\}_{k=1}^\infty$ in H. The coefficients will then be given by the Fourier coefficients formula: $u_k(t) = \langle u(t), \phi_k \rangle_H$.

We can similarly expand the source function:

$$f(t) = \sum_{k \in I} f_k(t) \phi_k, \quad f_k(t) = \langle f(t), \phi_k \rangle_H. \tag{3.7}$$

From (3.6) we have $\mathcal{L}\phi_k = \lambda_k \phi_k, \quad k \in I$. Hence,

$$\mathcal{L}u(t) = \sum_{k \in I} u_k(t) \lambda_k \phi_k \tag{3.8}$$

and

$$D_q u(t) = \sum_{k \in I} D_q u_k(t) \phi_k. \tag{3.9}$$

Substituting (3.8) and (3.9) into Eq. (3.1), we find

$$\sum_{k \in I} \left[D_q u_k(t) + \lambda_k \upsilon(t) u_k(t) \right] \phi_k = \sum_{k \in I} f_k(t) \phi_k. \tag{3.10}$$

But then, due to the completeness,

$$D_q u_k(t) + \lambda_k \upsilon(t) u_k(t) = f_k(t), \quad k \in I, \tag{3.11}$$

which are ODEs for the coefficients $u_k(t)$ of the series (3.6). Using the integrating factor $\gamma_\upsilon^{-1}(\lambda_k t)$ and (2.2) and (2.11), we can rewrite this equation as

$$\begin{aligned}\gamma_\upsilon^{-1}(q\lambda_k t) f_k(t) &= \gamma_\upsilon^{-1}(q\lambda_k t) D_q u_k(t) + \gamma_\upsilon^{-1}(q\lambda_k t)\lambda_k \upsilon(t) u_k(t) \\ &= \gamma_\upsilon^{-1}(q\lambda_k t) D_q u_k(t) + D_q\left[\gamma_\upsilon^{-1}(\lambda_k t)\right] u_k(t) \\ &= D_q\left[\gamma_\upsilon^{-1}(\lambda_k t) u_k(t)\right]. \end{aligned} \tag{3.12}$$

From (2.4) and (3.10) we get

$$\begin{aligned}&\int_0^t D_q\left[\gamma_\upsilon^{-1}(\lambda_k s) u_k(s)\right] d_q s \\ &= \int_0^t \gamma_\upsilon^{-1}(q\lambda_k s) f_k(s) d_q s \\ &\Longrightarrow \gamma_\upsilon^{-1}(\lambda_k t) u_k(t) = u_k(0) + \int_0^t \gamma_\upsilon^{-1}(q\lambda_k s) f_k(s) d_q s \\ &\Longrightarrow u_k(t) = \gamma_\upsilon(\lambda_k t) u_k(0) + \gamma_\upsilon(\lambda_k t)\int_0^t \gamma_\upsilon^{-1}(q\lambda_k s) f_k(s) d_q s.\end{aligned}$$

But the initial conditions (3.2) and (3.6) imply that

$$u(0) = \sum_{k\in I} u_k(0)\phi_k = \varphi \quad \Rightarrow \quad u_k(0) = \langle \varphi, \phi_k\rangle_H.$$

Thus,

$$u_k(t) = \langle \varphi, \phi_k\rangle_H \gamma_\upsilon(\lambda_k t) + \gamma_\upsilon(\lambda_k t)\int_0^t \gamma_\upsilon^{-1}(q\lambda_k s)\langle f(s), \phi_k\rangle_H d_q s. \tag{3.13}$$

So, the solution can be written in series form as

$$u(t) = \sum_{k\in I}\left[\langle \varphi, \phi_k\rangle_H \gamma_\upsilon(\lambda_k t) + \gamma_\upsilon(\lambda_k t)\int_0^t \gamma_\upsilon^{-1}(q\lambda_k s)\langle f(s), \phi_k\rangle_H d_q s\right]\phi_k,$$

giving (3.4).

Convergence By using Assumption 2.1 and (2.8) and (2.10), we obtain that

$$\frac{\gamma_\upsilon(\lambda_k t)}{\gamma_\upsilon(\lambda_k s)} \overset{(2.7)}{\leq} \frac{E_q^{\beta\lambda_k s}}{E_q^{\alpha\lambda_k t}} \leq \frac{E_q^{\alpha\lambda_k s}}{E_q^{\alpha\lambda_k t}} \leq 1, \quad 0 < s \leq t,$$

since $E_q^{\alpha\lambda_k t}$ is an increasing function for $t > 0$. Indeed, using (2.8) we see that

$$\frac{d}{dt}\left[E_q^{\alpha\lambda_k t}\right] = \sum_{m=1}^{\infty} q^{m(m-1)/2} [\alpha\lambda_k]^k \frac{m}{[m]_q} t^{m-1} \geq 0$$

for $t > 0$.

Hence, from (2.10) and (3.13) it follows that

$$\begin{aligned} |\langle u(t), \phi_k\rangle_H| &\overset{(3.13)}{\leq} \gamma_\upsilon(\lambda_k t)\, |\langle \varphi, \phi_k\rangle_H| + \int_0^t \frac{\gamma_\upsilon(\lambda_k t)}{\gamma_\upsilon(\lambda_k s)} |\langle f(s), \phi_k\rangle_H|\, d_q s \\ &\leq \frac{1}{E_q^{\alpha t}} |\langle \varphi, \phi_k\rangle_H| + \int_0^t |\langle f(s), \phi_k\rangle_H|\, d_q s \\ &\leq \max\left\{\frac{1}{E_q^{\alpha t}}, 1\right\}\left[|\langle \varphi, \phi_k\rangle_H| + \int_0^t |\langle f(s), \phi_k\rangle_H|\, d_q s\right], \end{aligned} \tag{3.14}$$

and using (3.11), we get that

$$\begin{aligned} \lambda_k^{d/2}\left|D_q u_k(t)\right| &\overset{(3.11)}{\leq} \lambda_k^{d/2+1} \upsilon(t)\, |\langle \varphi, \phi_k\rangle_H| + \lambda_k^{d/2} |\langle f(t), \phi_k\rangle_H| \\ &\leq \beta \left|\langle \lambda_k^{d/2+1} \varphi, \phi_k\rangle_H\right| + \lambda_k^{-1} \left|\langle \lambda_k^{d/2+1} f(t), \phi_k\rangle_H\right| \\ &\leq \beta \left|\langle \mathcal{L}^{d/2+1} \varphi, \phi_k\rangle_H\right| + \lambda_0^{-1} \left|\langle \mathcal{L}^{d/2+1} f(s), \phi_k\rangle_H\right| \\ &\leq \max\{\beta, \lambda_0^{-1}\}\left[\left|\langle \mathcal{L}^{d/2+1} \varphi, \phi_k\rangle_H\right| + \sup_{0\leq s\leq T} \left|\langle \mathcal{L}^{d/2+1} f(s), \phi_k\rangle_H\right|\right] \end{aligned} \tag{3.15}$$

and

$$\begin{aligned} \lambda_k^{d/2} |\langle \mathcal{L}u(t), \phi_k\rangle_H| &= \lambda_k^{d/2} |\langle \lambda_k u_k(t), \phi_k\rangle_H| \\ &\overset{(3.14)}{\lesssim} \left|\langle \lambda_k^{d/2+1} \varphi, \phi_k\rangle_H\right| + \int_0^t \left|\langle \lambda_k^{d/2+1} f(s), \phi_k\rangle_H\right| d_q s \end{aligned}$$

$$= \left|\langle \mathcal{L}^{d/2+1}\varphi, \phi_k\rangle_H\right| + \int_0^t \left|\langle \mathcal{L}^{d/2+1} f(s), \phi_k\rangle_H\right| d_q s$$

$$\leq \left|\langle \mathcal{L}^{d/2+1}\varphi, \phi_k\rangle_H\right| + T \sup_{0\leq s\leq T} \left|\langle \mathcal{L}^{d/2+1} f(s), \phi_k\rangle_H\right|$$

$$\leq \max\{1, T\} \left[\left|\langle \mathcal{L}^{d/2+1}\varphi, \phi_k\rangle_H\right| + \sup_{0\leq s\leq T} \left|\langle \mathcal{L}^{d/2+1} f(s), \phi_k\rangle_H\right|\right] \quad (3.16)$$

Since $\varphi \in \mathcal{H}^{d+2}_{\mathcal{L}}$, $f \in L^\infty\left([0, T]; \mathcal{H}^{d+2}_{\mathcal{L}}\right)$, and using the Plancherel identity, we have

$$\begin{aligned}
\|D_q u(t)\|^2_{\mathcal{H}^d_{\mathcal{L}}} &= \|\mathcal{L}^{\frac{d}{2}} D_q u(t)\|^2_H \\
&= \sum_{k\in I} \lambda_k^d \left|D_q u_k(t)\right|^2 \\
&\overset{(3.15)}{\lesssim} \sum_{k\in I} \left|\langle \mathcal{L}^{d/2+1}\varphi, \phi_k\rangle_H\right|^2 + \sup_{0\leq s\leq T} \sum_{k\in I} \left|\langle \mathcal{L}^{d/2+1} f(s), \phi_k\rangle_H\right|^2 \\
&= \|\varphi\|^2_{\mathcal{H}^{d+2}_{\mathcal{L}}} + \|f\|^2_{L^\infty\left([0,T];\mathcal{H}^{d+2}_{\mathcal{L}}\right)} < \infty,
\end{aligned}$$

and

$$\begin{aligned}
\|\mathcal{L}u(t)\|^2_{\mathcal{H}^d_{\mathcal{L}}} &= \|\mathcal{L}^{d/2}\mathcal{L}u(t)\|^2_H \\
&= \sum_{k\in I} \lambda_k^d \left|\langle \mathcal{L}u(t), \phi_k\rangle_H\right|^2 \\
&\overset{(3.16)}{\lesssim} \sum_{k\in I} \left|\langle \mathcal{L}^{d/2+1}\varphi, \phi_k\rangle_H\right|^2 + \sup_{0\leq s\leq T} \sum_{k\in I} \left|\langle \mathcal{L}^{d/2+1} f(s), \phi_k\rangle_H\right|^2 \\
&\leq \|\varphi\|^2_{\mathcal{H}^{d+2}_{\mathcal{L}}} + \|f\|^2_{L^\infty\left([0,T];\mathcal{H}^{d+2}_{\mathcal{L}}\right)} < \infty.
\end{aligned}$$

Hence, the above estimates imply that

$$\|\mathcal{L}u(t)\|^2_{\mathcal{H}^d_{\mathcal{L}}} + \|D_q u(t)\|^2_{\mathcal{H}^d_{\mathcal{L}}} \lesssim \|\varphi\|^2_{\mathcal{H}^{d+2}_{\mathcal{L}}} + \|f\|^2_{L^\infty\left([0,T];\mathcal{H}^{d+2}_{\mathcal{L}}\right)} < \infty,$$

which means that $u \in C^1_q\left([0, T]; \mathcal{H}^{d+2}_{\mathcal{L}}\right) \cap L^\infty\left([0, T]; \mathcal{H}^{d+2}_{\mathcal{L}}\right)$ and this yields the estimate (3.5).

Uniqueness It only remains to prove the uniqueness of the solution. We assume the opposite, namely, that there exist functions $u(t)$ and $v(t)$, which are two different solutions of Problem (3.1)–(3.2). Thus, we have that

$$\begin{cases} D_{q,t}u(t) + \varphi(t)\mathcal{L}u(t) = f(t), \ t > 0, \\ u(0) = \varphi \in H, \end{cases}$$

and

$$\begin{cases} D_{q,t}v(t) + \varphi(t)\mathcal{L}v(t) = f(t), \ t > 0, \\ v(0) = \varphi \in H. \end{cases}$$

We define $W(t) = u(t) - v(t)$. Then the function $W(t)$ is a solution of the following problem:

$$\begin{cases} D_q w(t) + \varphi(t)\mathcal{L}w(t) = 0, \ t > 0, \\ w(0, x) = 0. \end{cases}$$

From (3.6), it follows that $W(t) \equiv 0$, that is, $u(t) \equiv v(t)$, and this contradiction to our assumption proves the uniqueness of the solution. The proof is complete. □

References

1. V. Bardek, S. Meljanac, Deformed Heisenberg algebras, a Fock-space representation and the Calogero model. Eur. Phys. J. C-Part. Fields **17**, 539–547 (2000)
2. R. Hinterding, J. Wess, q-deformed Hermite polynomials in q-quantum mechanics. Eur. Phys. J. C-Part. Fields **6**, 183–186 (1999)
3. M. Micu, A q-deformed Schrodinger equation. J. Phys. A: Math. Gen. **32**, 77–65 (1999)
4. A. Lavagno, Basic-deformed quantum mechanics. Rep. Math. Phys. **64**, 79–91 (2009)
5. V.V. Eremin, A.A. Meldianov, The q-deformed harmonic oscillator, coherent states, and the uncertainty relation. Theor. Math. Phys. **147**, 709–715 (2006)
6. F.H. Jackson, On q-functions and a certain difference operator. Trans. Roy. Soc. Edin. **46**, 253–281 (1908)
7. F.H. Jackson, On a q-definite integrals. Q. J. Pure Appl. Math. **41**, 193–203 (1910)
8. R.D. Carmichael, The general theory of linear q-difference equations. Am. J. Math. **34**, 147–168 (1912)
9. P. Cheung, V. Kac, *Quantum Calculus* (Edwards Brothers, Inc., Ann Arbor, MI, 2000)
10. T. Ernst, *A Comprehensive Treatment of q-Calculus* (Birkhäuser/Springer, Basel AG, Basel, 2012)
11. M. Ruzhansky, S. Shaimardan, Direct and inverse source problems for heat equation in quantum calculus. https://doi.org/10.48550/arXiv.2212.07403

Evolutionary PDEs with Volterra Operators: Direct and Inverse Source Problems

Marian Slodička

Abstract Integro-differential problems arise in many practical situations when the solution of the governing PDE depends on its history. This is usually modelled by a Volterra operator. Such operators include also time-fractional derivatives, which are very in fashion in the last decades. Variable order fractional calculus represents a new direction in the research. There are still many open questions in this area, because the use of known techniques is very restricted. We try to point out the main difficulties, and we also state some open problems.

1 Introduction

Mathematical modelling of integro-differential evolutionary processes combines the governing time-dependent partial differential equation (PDE) with an ordinary differential equation (ODE) describing, e.g. the reaction, sedimentation, etc. In this way a kind of history effect is incorporated. Solvability of such problems has its roots at the theory of integral (Volterra and Fredholm) equations combined with the classical results from the PDEs.

Studies of such evolutionary PDEs with memory started in the 1970s of the twentieth century (cf., e.g. [1–3]). Some of the first papers solving parabolic integro-differential settings by semi-discretization in time (Rothe's) method can be found in [4–6]. The same proof-technique was applied to parabolic-hyperbolic systems with general Volterra operators, cf. [7]. First, we recall the definition of a Volterra operator.

M. Slodička (✉)
Department of Electronics and Information Systems, Research Group of Numerical Analysis and Mathematical Modeling (NaM^2), Ghent University, Gent, Belgium
e-mail: marian.slodicka@ugent.be

M. Ruzhansky, B. Torebek (eds.), *Extended Abstracts MWCAPDE 2023*,
Research Perspectives Ghent Analysis and PDE Center 1,
https://doi.org/10.1007/978-3-031-41665-1_23

Definition 1.1 (see [8]) The operator $E : L^\infty\left([0,T], L^2(\Omega)\right) \to L^\infty\left([0,T], L^2(\Omega)\right)$ is said to be a Volterra operator in $L^2(\Omega)$ if

$$\{u(s) = v(s) \text{ a.e. in } S_t = [0,t], t \in [0,T]\}$$
$$\Longrightarrow \{E(u)(s) = E(v)(s) \text{ a.e. in } S_t = [0,t], t \in [0,T]\}.$$

Here are a few examples $E(u)(t) = f(u(t)), \int_0^t K(t,s)u(s)\,\mathrm{d}s$, $f\left(\int_0^t K(t,s)u(s)\,\mathrm{d}s\right)$. The papers [4, 7] considered Volterra operators with bounded integral kernels, because the following classical Grönwall lemma was applied during the proofs.

Lemma 1.2 (Grönwall) *Let $r(t), h(t), y(t)$ be continuous real functions defined on $[a,b]$ satisfying $r(t), h(t) \geq 0$. If*

$$y(t) \leq h(t) + \int_a^t r(s)y(s)\,ds \qquad \textit{for} \qquad a \leq t \leq b,$$

then

$$y(t) \leq h(t) + \int_a^t h(s)r(s)\exp\left(\int_s^t r(\tau)d\tau\right) ds$$

is valid for all $t \in [a,b]$.

A lot of mathematical models contain weakly singular integral kernels. They can be handled in two different ways. The first path relies on the positive-definiteness of the singular integral kernel. So, singular term stays on the left-hand side by establishing energy estimates due to its non-negativity; see, e.g. [9, 10]. The key point is the following property of convolution kernel $g \in L^1_{\text{loc}}(0,\infty)$. If g obeys

$$(-1)^j g^{(j)}(t) \geqslant 0, \quad \forall t \geqslant 0, j = 0,1,2; \quad g' \not\equiv 0, \tag{1.1}$$

then it is strongly positive definite by [11, Corollary 2.2], i.e.

$$\int_0^t (g * u)(s)u(s)\,\mathrm{d}s \geqslant 0, \qquad t > 0, \quad \forall u \in L^2_{\text{loc}}((0,\infty)). \tag{1.2}$$

The symbol $*$ stands for the convolution on the positive half-line, i.e.

$$(k * v)(t) = \int_0^t k(t-s)v(s)\,\mathrm{d}s.$$

By the second manner, the singular term moves to the right-hand-side, and it has to be estimated by means of a generalized Grönwall lemma, cf. [12, Lemma 7.1.1],[13, Lemma 1]. For a discrete version of this lemma, we refer the reader to [14, 15].

Lemma 1.3 (generalized Grönwall, cf. [13, Lemma 1]) *Let a, w be non-negative continue real functions obeying*

$$w(t) \leq a(t) + C \int_0^t (t-s)^{\alpha-1} w(s)\, ds, \tag{1.3}$$

for $0 < \alpha \leq 1$, $C > 0$, $t \in [0, T]$. *Then*

$$w(t) \leq C \left(a(t) + \int_0^t (t-s)^{\alpha-1} a(s)\, ds \right), \qquad C = C(\alpha, T).$$

2 Fractional Calculus

Very important examples of Volterra operators can be found in fractional calculus (FC), including differentiation and/or integration (of an arbitrary real order). The first documented discussion on fractional-order differentials are letters exchanged between Leibniz and de l'Hôpital in 1695. Machado et al. [16] summarized historical perspective on the major developments in fractional calculus since the 1970s. Most studies in the literature concern the constant order (CO) fractional operators. There are several types of CO fractional derivatives (Caputo, Riemann-Liouville, Grünwald-Letnikov, etc.); see, e.g. [17]. Let us recall the definition of the Riemann-Liouville kernel $g_{1-\beta}$ given as

$$g_{1-\beta}(t) = \frac{t^{-\beta}}{\Gamma(1-\beta)}, \qquad t > 0,\ 0 < \beta < 1, \tag{2.1}$$

which is strongly positive definite.The Riemann-Liouville fractional integral is introduced as

$$\left(I^{\beta} u\right)(t) = \left(g_{1-\beta} * u\right)(t)$$

and the Caputo fractional derivative (cf., e.g. [17, 18]) reads as

$$\left(\partial_t^{\beta} u\right)(t) = \left(g_{1-\beta} * \partial_t u\right)(t).$$

The typical time fractional diffusion PDE with a general second-order differential operator A takes the following form:

$$\left(\partial_t^{\beta} u(x)\right)(t) + Au(x,t) = F(x,t).$$

This form can be derived as the diffusion limit of a continuous time random walk under the assumption that the mean waiting time has a power-law decaying tail [19, 20]. We would like to mention that the solution exhibits a nonphysical initial weak singularity for small $t > 0$. The mobile-immobile time fractional PDE

$$\partial_t u(x,t) + k\left(\partial_t^{\beta} u(x)\right)(t) + Au(x,t) = F(x,t), \tag{2.2}$$

was presented in [21]. Here we see that the left-hand side contains $\partial_t u$, which substantially simplifies the stability analysis. This is due to the fact that the fractional derivative can be incorporated in the right-hand side and handled by the generalized Grönwall Lemma 1.3. The solvability for the parabolic case was addressed in [22] and for the hyperbolic situation

$$\partial_{tt} u(x,t) + k\left(\partial_t^{\beta} u(x)\right)(t) + Au(x,t) = F(x,t), \tag{2.3}$$

in [23]. Let us note that [22, 23] considered just linear PDEs. Semilinear versions of both (parabolic and hyperbolic) cases were studied in [24].

The author of [24] considered a (nonlinear) Volterra operator $E(v)(t)$ obeying for $t \in [0, T]$

$$|E(u)(t) - E(v)(t)| \le C\sqrt{\int_0^t (t-s)^{-\overline{\beta}}(u(s)-v(s))^2 \, \mathrm{d}s}, \quad 0 \le \overline{\beta} < 1. \tag{2.4}$$

Interesting examples of such operators include also the so-called variable-order (VO) fractional derivatives, cf. [24]. Inspecting the literature one can find several definitions of VO-fractional derivatives. One can define the time-dependent version of the Riemann-Liouville kernel (2.1) as follows:

$$\left(g_{1-\beta(t)}\right)(t) = \frac{t^{-\beta(t)}}{\Gamma(1-\beta(t))} \qquad t > 0,\ 0 < \beta(t) < 1. \tag{2.5}$$

Using this, the suitable form of VO Riemann-Liouville integral reads as

$$\left(I^{\beta(t)} u\right)(t) = \left(g_{1-\beta(t)} * u\right)(t) = \int_0^t \frac{(t-s)^{-\beta(t-s)}}{\Gamma(1-\beta(t-s))} u(s) \, \mathrm{d}s$$
$$t > 0,\ 0 < \beta(t) < 1 \tag{2.6}$$

and the Caputo VO fractional derivative is given by

$$\left(\partial_t^{\beta(t)} u\right)(t) = \left(g_{1-\beta(t)} * \partial_t u\right)(t) = \int_0^t \frac{(t-s)^{-\beta(t-s)}}{\Gamma(1-\beta(t-s))} \partial_t u(s)\, \mathrm{d}s$$
$$t > 0,\ 0 < \beta(t) < 1. \tag{2.7}$$

An another form of VO-derivative reads as

$$\left(\partial_t^{\beta(t)} u(x)\right)(t) = \frac{1}{\Gamma(1-\beta(t))} \int_0^t (t-s)^{-\beta(t)} u_t(x,s)\, \mathrm{d}s. \tag{2.8}$$

2.1 Direct Parabolic Problem

Please note that both possibilities (2.7) and (2.8) represent a Volterra operator obeying (2.4), cf. [24]. We would like to point out that the paper [24] also presented a very interesting technique based on Banach contraction principle. This has been applied to show the solvability of the following direct initial boundary-value problem

$$\begin{aligned} u_t(t) - \Delta u(t) &= F(t) + E(u, \nabla u, u_t)(t) && \text{in } \Omega \times (0,T) \\ u(t) &= 0 && \text{in } \Gamma \times [0,T] \\ u(0) &= u_0 && \text{in } \Omega \end{aligned} \tag{2.9}$$

with a Volterra operator E satisfying (2.4). Introducing the Banach space

$$\mathbb{X}_\eta := \left\{ v : [0,\eta] \to H_0^1(\Omega);\ \max_{t\in[0,\eta]} \|\nabla v(t)\|^2 + \int_0^\eta \|v_t(t)\|^2 \,\mathrm{d}t < \infty \right\}$$

one can show that

$$u \in \mathbb{X}_\eta \Longrightarrow E(u, \nabla u, u_t)(t) \in L^2\left((0,\eta), L^2(\Omega)\right).$$

Let us consider $F \in L^2\left((0,\eta), L^2(\Omega)\right)$ and an auxiliary problem

$$\begin{aligned} v_t(t) - \Delta v(t) &= F(t) + E(u, \nabla u, u_t)(t) && \text{in } \Omega \times (0,T) \\ v(t) &= 0 && \text{in } \Gamma \times [0,T] \\ v(0) &= u_0 && \text{in } \Omega \end{aligned} \tag{2.10}$$

The theory of linear parabolic equations implies existence of a unique solution $v \in \mathbb{X}_\eta$. This naturally defines a map

$$A : \mathbb{X}_\eta \to \mathbb{X}_\eta, \qquad Au = v.$$

It can be proved that A is a contraction on $\mathbb{X}_\eta$ provided η is sufficiently small. The contraction mapping theorem implies that there exists a unique $u \in \mathbb{X}_\eta$ such that $Au = u$. It turns out that we can extend the solution beyond the time η by a bootstrap argument.

2.2 Parabolic Inverse Source Problem

Consider (2.9) with a separated source $F(x,t) = f(x)h(t)$, where $f(x)$ is given and the unknown function $h(t)$ has to be recovered from an additional local measurement

$$m(t) = \int_\Omega \omega(x) u(x,t)\,\mathrm{d}x. \tag{2.11}$$

One can get rid of the unknown $h(t)$ as follows. Multiplying the PDE by $\omega(x)$ and integrating over Ω bring us to

$$m't) - (\Delta u, \omega) = h(t)\,(f, \omega) + (E(u, \nabla u, u_t)(t), \omega)\,.$$

Involving the Green argument, we may write (if $(f, \omega) \neq 0$)

$$h(t) = \frac{1}{(f,\omega)} \left(m'(t) + (\nabla u, \nabla \omega) - (E(u, \nabla u, u_t)(t), \omega) \right). \tag{2.12}$$

Wrapping the considerations above, we see that the inverse source problem – to find $\{u(x,t), h(t)\}$ obeying (2.9) with a separated source $F(x,t) = f(x)h(t)$ along with (2.11) – can be rewritten as a direct problem:

$$\begin{aligned}
u_t(t) & -\Delta u(t) \\
& = E(u, \nabla u, u_t)(t) \\
& + \frac{f}{(f,\omega)} \left(m'(t) + (\nabla u(t), \nabla \omega) - (E(u, \nabla u, u_t)(t), \omega) \right) && \text{in } \Omega \times (0,T) \\
u(t) & = 0 && \text{in } \Gamma \times [0,T] \\
u(0) & = u_0 && \text{in } \Omega
\end{aligned} \tag{2.13}$$

Solvability can be addressed using the contraction mapping theorem.

2.3 Hyperbolic Direct Problem

Similar proof-technique based on contraction mapping theorem is applicable to a hyperbolic case

$$\begin{aligned} u_{tt}(t) - \Delta u(t) &= F(t) + E(u, \nabla u, u_t)(t) && \text{in } \Omega \times (0, T) \\ u(t) &= 0 && \text{in } \Gamma \times [0, T] \\ u_t(0) &= w_0 && \text{in } \Omega \\ u(0) &= u_0 && \text{in } \Omega, \end{aligned} \tag{2.14}$$

Here, we have to work in a different Banach space:

$$\mathbb{Y}_\eta := \left\{ v : [0, \eta] \to H_0^1(\Omega);\ \max_{t \in [0,\eta]} \|\nabla v(t)\|^2 + \max_{t \in [0,\eta]} \|v_t(t)\|^2 < \infty \right\}$$

Theorem 2.1 (See [24]) *Assume (2.4), $w_0 \in L^2(\Omega)$, $u_0 \in H_0^1(\Omega)$, $F \in L^2\left((0, T), L^2(\Omega)\right)$. Then the problem (2.14) admits a unique weak solution $u \in \mathbb{Y}_T$.*

2.4 Hyperbolic Inverse Source Setting

Let us consider (2.14) with a separated source of the form $F(x, t) = f(x)h(t)$. Here $f(x)$ is given, and the unknown function $h(t)$ has to be recovered from an additional measurement (2.11). Measuring the governing PDE from (2.14), we get

$$m''t) - (\Delta u, \omega) = h(t)\,(f, \omega) + (E(u, \nabla u, u_t)(t), \omega)\,.$$

Applying the Green argument, we can write (if $(f, \omega) \neq 0$)

$$h(t) = \frac{1}{(f, \omega)} \left(m''(t) + (\nabla u, \nabla \omega) - (E(u, \nabla u, u_t)(t), \omega) \right). \tag{2.15}$$

Based on this, we see that the inverse source problem – to find $\{u(x, t), h(t)\}$ obeying (2.14) with a separated source $F(x, t) = f(x)h(t)$ along with (2.11) – can be reformulated as a direct problem:

$$\begin{aligned} u_{tt}(t) & -\Delta u(t) \\ &= E(u, \nabla u, u_t)(t) \\ &+ \frac{f}{(f, \omega)} \left(m''(t) + (\nabla u(t), \nabla \omega) - (E(u, \nabla u, u_t)(t), \omega) \right) && \text{in } \Omega \times (0, T) \\ u(t) &= 0 && \text{in } \Gamma \times [0, T] \\ u_t(0) &= w_0 && \text{in } \Omega \\ u(0) &= u_0 && \text{in } \Omega. \end{aligned} \tag{2.16}$$

The well-posedness of the inverse source problem reads as follows.

Theorem 2.2 (See [24]) *Consider the inverse source problem – to find* $\{u(x,t), h(t)\}$ *obeying (2.14) with a separated source* $F(x,t) = f(x)h(t)$ *along with (2.11). Assume (2.4),* $u_0 \in L^2(\Omega)$, $w_0 \in H_0^1(\Omega)$, $f \in L^2(\Omega)$, $m'' \in L^2((0,T))$, $\omega \in H_0^1(\Omega)$, $(f,\omega) \neq 0$. *Then this setting admits a unique weak solution* $u \in \mathbb{Y}_T$ *and* $h \in L^2((0,T))$.

3 Some Open Problems for VO Time-Fractional Derivatives

We would like to point out that the attractive technique from [24]] works just in the situation that the differential operator on the left-hand side of the PDE is of a parabolic or hyperbolic form. When we replace the whole derivatives u_t or u_{tt} by the fractional ones with a time-dependent order, big troubles arise. Kochubei [25, 26] developed general fractional calculus for

$$D_\phi f(t) := \frac{d}{dt}(\phi * f)(t) - \phi(t) f(0),$$

which is based on Laplace transform. The integral kernel has to obey some conditions constraining ϕ to be completely monotone (cf. [27]), which is very restrictive. The paper [27] shows a few interesting examples (exponential, Mittag-Leffler, and error function-type transitions), which could be of practical relevance.

We recall that the standard approach for handling CO time-fractional derivative β is based on positive definiteness of $g_{1-\beta}$ – see (1.2). This way cannot be applied for $\beta(t)$ from (2.6) due to the lack of positive definiteness of $g_{1-\beta(t)}$. This can be easily seen by the following deduction. The kernel $g_{1-\beta(t)}$ keeps its sign, namely, $\left(g_{1-\beta(t)}\right)(t) \geq 0$. For its derivative, we can write

$$\begin{aligned}
\left(g_{1-\beta(t)}(t)\right)' &= \left(\frac{t^{-\beta(t)}}{\Gamma(1-\beta(t))}\right)' = \left(e^{-\beta(t)\ln t - \ln(\Gamma(1-\beta(t)))}\right)' \\
&= \left(\frac{t^{-\beta(t)}}{\Gamma(1-\beta(t))}\right)\left[-\beta'(t)\ln t - \frac{\beta(t)}{t} + \frac{\Gamma'(1-\beta(t))}{\Gamma(1-\beta(t))}\beta'(t)\right].
\end{aligned}$$

This can change its sign.

Finally, we formulate some open problems concerning the solvability of

$$\left(\partial_t^{\beta(t)} u\right)(t) + Au = f(u).$$

What to do in these situations? Can we address the well-posedness?

- $\beta(t)$ is smooth
- $\beta(t)$ has jumps
- Which forms can $\beta(t)$ take?

- Can $\beta(t)$ reach 1? $0 < \beta(t) \leq 1$?
- What if $1 < \beta(t) < 2$?
- What to do with inverse source problems by determination of $f(x)$ or $h(t)$, when the source term has the form $F(x, t) = f(x)h(t)$?

References

1. C.M. Dafermos, An abstract Volterra equation with applications to linear viscoelasticity. J. Differ. Equ. **7**, 554–569 (1970)
2. R.C. MacCamy, J.S.W. Wong, Stability theorems for some functional equations. Trans. Am. Math. Soc. **164**, 1–37 (1972)
3. R.K. Miller, An integrodifferential equation for rigid heat conductors with memory. J. Math. Anal. Applic. **66**(2), 313–332 (1978)
4. M. Slodička, Parabolic partial differential equations with memory. Math. Slovaca **34**(1), 3–23 (1984)
5. I.H. Sloan, V. Thomée, Time discretization of an integro-differential equation of parabolic type. SIAM J. Numer. Anal. **23**(5), 1052–1061 (1986)
6. J. Kačur, Application of Rothe's method to evolution integrodifferential equations. J. Reine Angew. Math. **1988**(388), 73–105 (1988)
7. M. Slodička, Application of Rothe's method to evolution integrodifferential systems. Comment. Math. Univ. Carol. **30**(1), 57–70 (1989)
8. H. Gajewski, K. Gröger, K. Zacharias, Nichtlineare Operatorgleichungen und Operatordifferentialgleichungen, in *Mathematische Lehrbücher und Monographien. II. Abteilung*, Band 38 (Akademie, Berlin, 1974)
9. O.J. Staffans, Nonlinear Volterra integral equations with positive definite kernels. Proc. Am. Math. Soc. **51**(1), 103–108 (1975)
10. J.S.W. Wong, Positive definite functions and Volterra integral equations. Bull. Am. Math. Soc. **80**, 679–682 (1974)
11. J.A. Nohel, D.F. Shea, Frequency domain methods for Volterra equations. Adv. Math. **22**(3), 278–304 (1976)
12. D. Henry, Geometric theory of semilinear parabolic equations, in *Lecture Notes in Mathematics*, vol. 840 (Springer, Berlin-Heidelberg-New York, 1981)
13. C. Chen, V. Thomée, L.B. Wahlbin, Finite element approximation of a parabolic integro-differential equation with a weakly singular kernel. Math. Comput. **58**, 587–602 (1992)
14. M. Slodička, Semigroup formulation of Rothe's method: application to parabolic problems. Comment. Math. Univ. Carol. **33**(2), 245–260 (1992)
15. C.M. Elliott, S. Larsson, Error estimates with smooth and nonsmooth data for a finite element method for the cahn-hilliard equation. Math. Comput. **58**, 603–630 (1992)
16. J.T. Machado, V. Kiryakova, F. Mainardi, Recent history of fractional calculus. Commun. Nonlinear Sci. Numer. Simul. **16**(3), 1140–1153 (2011)
17. I. Podlubný, *Fractional Differential Equations: An Introduction to Fractional Derivatives, Fractional Differential Equations, to Methods of Their Solution and Some of Their Applications*. Mathematics in science and engineering (Academic Press, San Diego, 1999)
18. M. Caputo, Linear models of dissipation whose Q is almost frequency independent-II. Geophys. J. R. Astron. Soc. **13**(5), 529–539 (1967)
19. R. Metzler, J. Klafter, The random walk's guide to anomalous diffusion: a fractional dynamics approach. Phys. Rep. **339**, 1–77 (2000).
20. M.M. Meerschaert, A. Sikorskii, *Stochastic Models for Fractional Calculus*, (De Gruyter, Berlin, 2012). https://doi.org/10.1515/9783110258165

21. R. Schumer, D.A. Benson, M.M. Meerschaert, B. Baeumer, Fractal mobile/immobile solute transport. Water Resour. Res. **39**, 1296 (2003)
22. X. Zheng, H. Wang, Optimal-order error estimates of finite element approximations to variable-order time-fractional diffusion equations without regularity assumptions of the true solutions. IMA J. Numer. Anal. **41**(2), 1522–1545 (2020)
23. X. Zheng, H. Wang, Analysis and discretization of a variable-order fractional wave equation. Commun. Nonlinear Sci. Numer. Simul. **104**, 106047 (2022)
24. M. Slodička, Some direct and inverse source problems in nonlinear evolutionary PDEs with Volterra operators. Inverse Prob. **38**(12), 124001 (2022)
25. A.N. Kochubei, General fractional calculus, evolution equations, and renewal processes. Integr. Equ. Oper. Theory **71**(4), 583–600 (2011)
26. A.N. Kochubei, General fractional calculus. Handbook Fractional Calculus Applic. **1**, 111–126 (2019)
27. R. Garrappa, A. Giusti, F. Mainardi, Variable-order fractional calculus: a change of perspective. Commun. Nonlinear Sci. Numer. Simul. **102**, 105904 (2021). https://doi.org/10.1016/j.cnsns.2021.105904

Inverse Source Problems in Thermoelasticity

Karel Van Bockstal

Abstract In thermoelasticity, a parabolic equation for the temperature and a hyperbolic equation for the displacement of a body are coupled. In this research summary, an overview of the uniqueness results on inverse source problems in this setting is provided.

2020 Mathematics Subject Classification 35E15, 35A02, 74H75

1 Introduction

We consider an isotropic homogeneous thermoelastic body occupying an open and bounded domain $\Omega \subset \mathbb{R}^d$ with a Lipschitz continuous boundary Γ, where $d \in \{1, 2, 3\}$. Let $Q_T = \Omega \times (0, T)$ and $\Sigma_T = \Gamma \times (0, T)$, where $T > 0$ is a given final time. The coupled thermoelastic system of type III [1] is considered:

$$\begin{cases} \rho \partial_{tt}\mathbf{u} - \mu\Delta\mathbf{u} - (\lambda+\mu)\nabla(\nabla\cdot\mathbf{u}) + \gamma\nabla\theta = \mathbf{p} & \text{in } Q_T; \\ \rho C_s \partial_t\theta - \kappa\Delta\theta - k * \Delta\theta + T_0\gamma\nabla\cdot\partial_t\mathbf{u} = h & \text{in } Q_T; \\ \mathbf{u}(\cdot,0) = \overline{\mathbf{u}}_0(\cdot), \quad \partial_t\mathbf{u}(\cdot,0) = \overline{\mathbf{u}}_1(\cdot), \quad \theta(\cdot,0) = \overline{\theta}_0(\cdot) & \text{in } \Omega. \end{cases} \tag{1.1}$$

This system describes the elastic and thermal behaviours in Ω represented by the displacement $\mathbf{u}$ and the temperature difference θ from the reference value $T_0 > 0$ (the reference temperature corresponds to the undeformed and unstressed state of the material). The Lamé parameters λ and μ, the mass density ρ, the specific heat C_s, the coupling (absorbing) coefficient γ, and the thermal coefficient κ are assumed

K. Van Bockstal (✉)
Ghent Analysis & PDE Center, Department of Mathematics: Analysis, Logic and Discrete Mathematics, Ghent University, Ghent, Belgium
e-mail: karel.vanbockstal@ugent.be

M. Ruzhansky, B. Torebek (eds.), *Extended Abstracts MWCAPDE 2023*,
Research Perspectives Ghent Analysis and PDE Center 1,
https://doi.org/10.1007/978-3-031-41665-1_24

to be positive constants, whilst $\mathbf{p}$ and h denote the load (body force) and heat source, respectively. The symbol $*$ denotes the Laplace convolution product

$$(k * \theta)(\mathbf{x}, t) := \int_0^t k(t-s)\theta(\mathbf{x}, s)\mathrm{d}s, \qquad (\mathbf{x}, t) \in Q_T.$$

This term takes memory effects into account. A typical example for k is $k(t) = a\exp(-bt)$ with $a, b > 0$.

Goal of the paper In this paper, we summarise the results obtained in [2–6] about inverse source problems (ISPs) in thermoelasticity. The sources $\mathbf{p}$ and h are rewritten as

$$\mathbf{p}(\mathbf{x}, t) = g_{\mathbf{p}}(t)\mathbf{f}_{\mathbf{p}}(\mathbf{x}) + \tilde{\mathbf{p}}(\mathbf{x}, t),$$

$$h(\mathbf{x}, t) = g_h(t) f_h(\mathbf{x}) + \tilde{h}(\mathbf{x}, t),$$

where $\tilde{\mathbf{p}}$ and $\tilde{h}$ are the given part of the source. In the inverse source problems under consideration, either $g_h(t)$, $\mathbf{f}_{\mathbf{p}}(\mathbf{x})$ or $f_h(\mathbf{x})$ may be unknown.

Preliminary results about ISPs in thermoelasticity In [7], it was shown that the measurements $\mathbf{u}|_{\omega\times(0,T)}$ and $\theta(\cdot, t_0)$ uniquely determine the spatial component $f_h(\mathbf{x})$ in a heat source for type I thermoelastic system ($\kappa \neq 0$ and $k = 0$) of the form $h(\mathbf{x}, t) = f_h(\mathbf{x})\sigma(\mathbf{x}, t)$ with σ given. Here, $\omega \subset \Omega$ is a given subdomain satisfying $\partial\omega \supset \partial\Omega$, and $t_0 \in (0, T)$ is an appropriate given time. The article [8] provides a local uniqueness result for the recovery of a spatially varying source term $\mathbf{p}(\mathbf{x}, t) = \sigma(\mathbf{x}, t)\mathbf{f}_{\mathbf{p}}(\mathbf{x})$ in type II thermoelasticity ($\kappa = 0$ and $k \neq 0$), where σ is a known scalar function. Here $\mathbf{f}_{\mathbf{p}}(\mathbf{x})$ is determinable from the observation $\mathbf{u}|_{\omega\times(0,T)}$ provided that $\mathbf{f}_{\mathbf{p}}(\mathbf{x})$ is known in a neighbourhood ω_0 of $\partial\Omega$. Both Bellassoued and Yamamoto and Wu and Liu [7, 8] have considered $\Omega \subset \mathbb{R}^3$, homogeneous Dirichlet boundary conditions for temperature and displacement, and employed a Carleman estimate to obtain their results. No numerical scheme for computations is provided in these works.

Well-posedness direct problem The following well-posedness result for the direct problem can be obtained using Rothe's method; see, e.g. [4, Section 4]. Further, $(\cdot, \cdot)$ denotes the standard L^2-inner product, and $\langle \cdot, \cdot \rangle$ denotes the duality pairing.

Lemma 1.1 (Well-posedness forward problem) *Suppose that* $\boldsymbol{u} = \boldsymbol{0}$ *on* Γ *and* $\theta = 0$ *on* Γ. *Let the kernel* $k \in \mathrm{C}([0, T])$.

(i) *Assume that* $\boldsymbol{p} \in \mathrm{L}^2\left([0, T], \boldsymbol{L}^2(\Omega)\right)$, $h \in \mathrm{L}^2\left([0, T], \mathrm{L}^2(\Omega)\right)$, $\overline{\boldsymbol{u}}_0 \in \boldsymbol{H}_0^1(\Omega)$, $\overline{\boldsymbol{u}}_1 \in \boldsymbol{L}^2(\Omega)$ *and* $\overline{\theta}_0 \in \mathrm{L}^2(\Omega)$. *Then, the forward problem* (1.1) *has a unique weak solution* $\{\boldsymbol{u}, \theta\}$ *satisfying*

- $\boldsymbol{u} \in \mathrm{C}\left([0, T], \boldsymbol{L}^2(\Omega)\right) \cap \mathrm{L}^2\left((0, T), \boldsymbol{H}_0^1(\Omega)\right)$, $\partial_t \boldsymbol{u} \in \mathrm{L}^2\left((0, T), \boldsymbol{L}^2(\Omega)\right)$, $\partial_{tt}\boldsymbol{u} \in \mathrm{L}^2\left((0, T), \boldsymbol{H}_0^1(\Omega)^*\right)$;
- $\theta \in \mathrm{C}\left([0, T], \mathrm{L}^2(\Omega)\right) \cap \mathrm{L}^2\left((0, T), \mathrm{H}_0^1(\Omega)\right)$, $\partial_t\theta \in \mathrm{L}^2\left((0, T), \mathrm{H}_0^1(\Omega)^*\right)$,

such that

$$\rho\langle\partial_{tt}\boldsymbol{u}(t),\boldsymbol{\varphi}\rangle+\mu\,(\nabla\boldsymbol{u}(t),\nabla\boldsymbol{\varphi})$$
$$+\,(\lambda+\mu)\,(\nabla\cdot\boldsymbol{u}(t),\nabla\cdot\boldsymbol{\varphi})+\gamma\,(\nabla\theta(t),\boldsymbol{\varphi})=(\boldsymbol{p}(t),\boldsymbol{\varphi})$$

and

$$\rho C_s\langle\partial_t\theta(t),\psi\rangle+\kappa\,(\nabla\theta(t),\nabla\psi)$$
$$+\,((k*\nabla\theta)\,(t),\nabla\psi)-T_0\gamma\,(\partial_t\boldsymbol{u}(t),\nabla\psi)=(h(t),\psi)\,,$$

for all $\boldsymbol{\varphi}\in\boldsymbol{H}_0^1(\Omega)$, $\psi\in\mathrm{H}_0^1(\Omega)$ *and a.a.* $t\in(0,T]$.

(ii) *Assume that* $\boldsymbol{p}\in\mathrm{H}^1\left([0,T],\boldsymbol{L}^2(\Omega)\right)$, $h\in\mathrm{H}^1\left(([0,T],\mathrm{L}^2(\Omega)\right)$, $\overline{\boldsymbol{u}}_0\in\boldsymbol{H}^2(\Omega)\cap\boldsymbol{H}_0^1(\Omega)$, $\overline{\boldsymbol{u}}_1\in\boldsymbol{H}_0^1(\Omega)$ *and* $\overline{\theta}_0\in\mathrm{H}^2(\Omega)\cap\mathrm{H}_0^1(\Omega)$. *Let the PDEs be fulfilled for* $t=0$. *Then, the forward problem* (1.1) *has a unique weak solution* $\{\boldsymbol{u},\theta\}$ *satisfying*

- $\boldsymbol{u}\in\mathrm{C}\left([0,T],\boldsymbol{H}_0^1(\Omega)\right)$, $\partial_t\boldsymbol{u}\in\mathrm{C}\left([0,T],\boldsymbol{L}^2(\Omega)\right)\cap\mathrm{L}^2\left((0,T),\boldsymbol{H}_0^1(\Omega)\right)$, $\partial_{tt}\boldsymbol{u}\in\mathrm{L}^2\left((0,T),\mathrm{L}^2(\Omega)\right)$;
- $\theta\in\mathrm{C}\left([0,T],\mathrm{H}_0^1(\Omega)\right)$, $\partial_t\theta\in\mathrm{L}^2\left((0,T),\mathrm{H}_0^1(\Omega)\right)$,

such that

$$\rho\,(\partial_{tt}\boldsymbol{u}(t),\boldsymbol{\varphi})+\mu\,(\nabla\boldsymbol{u}(t),\nabla\boldsymbol{\varphi})$$
$$+\,(\lambda+\mu)\,(\nabla\cdot\boldsymbol{u}(t),\nabla\cdot\boldsymbol{\varphi})+\gamma\,(\nabla\theta(t),\boldsymbol{\varphi})=(\boldsymbol{p}(t),\boldsymbol{\varphi})$$

and

$$\rho C_s\,(\partial_t\theta(t),\psi)+\kappa\,(\nabla\theta(t),\nabla\psi)$$
$$+\,((k*\nabla\theta)\,(t),\nabla\psi)-T_0\gamma\,(\partial_t\boldsymbol{u}(t),\nabla\psi)=(h(t),\psi)\,,$$

for all $\boldsymbol{\varphi}\in\boldsymbol{H}_0^1(\Omega)$, $\psi\in\mathrm{H}_0^1(\Omega)$ *and a.a.* $t\in(0,T]$.

2 Identification of the Space-Dependent Sources $\mathbf{f_p}$ and f_h

First, we formulate the problem under consideration.

Problem 2.1 Can we find a unique $\mathbf{f_p}(\mathbf{x})$ and/or $f_h(\mathbf{x})$ from the additional final in time measurements

$$\mathbf{u}(\mathbf{x},T)=\boldsymbol{\xi}_T(\mathbf{x})\quad\text{and/or}\quad\theta(\mathbf{x},T)=\zeta_T(\mathbf{x})$$

or from the time-average measurements

$$\int_0^T \mathbf{u}(\mathbf{x}, t)\, \mathrm{d}t = \boldsymbol{\xi}_T(\mathbf{x}) \quad \text{and/or} \quad \int_0^T \theta(\mathbf{x}, t)\, \mathrm{d}t = \zeta_T(\mathbf{x}),$$

and can we provide a numerical scheme for computations?

The first result can be found in [2, Theorem 2.1].

Lemma 2.2 (Problem 2.1: answer I) *It is possible to recover $\boldsymbol{f_p}(\boldsymbol{x})$ uniquely for all types of thermoelasticity from the condition of final overdetermination*

$$\boldsymbol{u}(\boldsymbol{x}, T) = \boldsymbol{\xi}_T(\boldsymbol{x}), \quad \boldsymbol{\xi}_T \in \boldsymbol{L}^2(\Omega),$$

in the presence of a damping term $\boldsymbol{d}(\partial_t \boldsymbol{u})$ in the hyperbolic equation of the thermoelastic system, i.e.

$$\begin{cases} \rho\partial_{tt}\boldsymbol{u} + \boldsymbol{d}(\partial_t\boldsymbol{u}) - \mu\Delta\boldsymbol{u} - (\lambda+\mu)\nabla(\nabla\cdot\boldsymbol{u}) + \gamma\nabla\theta = \boldsymbol{f_p}(\boldsymbol{x}) & \text{in } Q_T; \\ \rho C_s\partial_t\theta - \kappa\Delta\theta - k * \Delta\theta + T_0\gamma\nabla\cdot\partial_t\boldsymbol{u} = 0 & \text{in } Q_T; \\ \boldsymbol{u} = \boldsymbol{0}, \quad \theta = 0 & \text{on } \Sigma_T; \\ \boldsymbol{u}(\cdot,0) = \partial_t\boldsymbol{u}(\cdot,0) = \boldsymbol{0}, \quad \theta(\cdot,0) = 0 & \text{in } \Omega. \end{cases}$$

It is assumed that:

- *The solution satisfies the properties of Lemma 1.1(ii); in particular it is necessary (by the variational approach) that $\partial_t \boldsymbol{u} \in \mathrm{L}^2\left((0,T), \boldsymbol{H}_0^1(\Omega)\right)$;*
- *The damping term $\mathbf{d} : \mathbb{R}^d \mapsto \mathbb{R}^d$ is component-wise strictly monotone increasing;*
- *The kernel $k \in \mathrm{C}^2([0,T])$ satisfies*

$$(-1)^j k^{(j)}(t) \geq 0 \quad \text{and} \quad k'(t) \not\equiv 0 \quad \forall t \in [0,T], \quad j = 0,1,2,$$

i.e. k is strongly positive definite [9, Corollary 2.2].

If $\mathbf{d}$ is linear (i.e. $\mathbf{d} = \bar{d}\mathbf{I}$ with $\bar{d} > 0$), then in [2, Section 3] a stable iterative algorithm is proposed to recover $\mathbf{f_p}$. It is an extension of the Landweber–Friedmann iteration scheme of [10] for the heat equation to thermoelastic systems but without the use of an adjoint problem. The results obtained in [2] are extended in [4] to anisotropic thermoelastic systems including kernels $k \in \mathrm{L}^1(0,T)$ satisfying $\int_0^T |k(t)|\,\mathrm{d}t \leq \kappa_0$, where κ_0 is the ellipticity constant of the thermal conductivity tensor (for isotropic thermoelasticity $\kappa_0 = \kappa$). In [2, 4], we have that $g_{\mathbf{p}} = 1$.

Recently, in [5], we have obtained uniqueness results to Problem 2.1 for a more general source. We summarise the results below.

Lemma 2.3 (Problem 2.1: answer II) *Consider the system*

$$\begin{cases} \rho\partial_{tt}\boldsymbol{u} - \mu\Delta\boldsymbol{u} - (\lambda+\mu)\nabla(\nabla\cdot\boldsymbol{u}) + \gamma\nabla\theta = g_p(t)\boldsymbol{f_p}(\boldsymbol{x}) & \text{in } Q_T; \\ \rho C_s\partial_t\theta - \kappa\Delta\theta - k*\Delta\theta + T_0\gamma\nabla\cdot\partial_t\boldsymbol{u} = 0 & \text{in } Q_T; \\ \boldsymbol{u} = \boldsymbol{0}, \quad \theta = 0 & \text{on } \Sigma_T; \\ \boldsymbol{u}(\cdot,0) = \partial_t\boldsymbol{u}(\cdot,0) = \boldsymbol{0}, \quad \theta(\cdot,0) = 0 & \text{in } \Omega. \end{cases}$$

It is possible to recover $\boldsymbol{f_p}(\boldsymbol{x})$ *uniquely for all types of thermoelasticity from*

$$\boldsymbol{u}(\boldsymbol{x},T) = \boldsymbol{\xi}_T(\boldsymbol{x}), \quad \boldsymbol{\xi}_T \in \boldsymbol{L}^2(\Omega),$$

assuming that $g_p \in \mathrm{C}^1([0,T])$ *satisfies either*

$$g_p(t) > 0 \quad \text{and} \quad g_p'(t) > 0, \quad \forall t \in [0,T]$$

or

$$g_p(t) < 0 \quad \text{and} \quad g_p'(t) < 0, \quad \forall t \in [0,T].$$

It is assumed that:

- *The solution satisfies the properties of Lemma 1.1(ii);*
- *The kernel k is strongly positive definite.*

This result can be found in [5, Theorem 2.2]. Moreover, the monotonicity conditions on g_p are the same as obtained in [11, Chapter 7] for the wave equation. Next, we state the content of [5, Theorem 2.7].

Lemma 2.4 (Problem 2.1: answer III) *Consider the system*

$$\begin{cases} \rho\partial_{tt}\boldsymbol{u} - \mu\Delta\boldsymbol{u} - (\lambda+\mu)\nabla(\nabla\cdot\boldsymbol{u}) + \gamma\nabla\theta = g_p(t)\boldsymbol{f_p}(\boldsymbol{x}) & \text{in } Q_T; \\ \rho C_s\partial_t\theta - \kappa\Delta\theta - k*\Delta\theta + T_0\gamma\nabla\cdot\partial_t\boldsymbol{u} = 0 & \text{in } Q_T; \\ \boldsymbol{u} = \boldsymbol{0}, \quad \theta = 0 & \text{on } \Sigma_T; \\ \boldsymbol{u}(\cdot,0) = \partial_t\boldsymbol{u}(\cdot,0) = \boldsymbol{0}, \quad \theta(\cdot,0) = 0 & \text{in } \Omega. \end{cases}$$

It is possible to recover $\boldsymbol{f_p}(\boldsymbol{x})$ *uniquely for all types of thermoelasticity from*

$$\int_0^T \boldsymbol{u}(\boldsymbol{x},t)\,\mathrm{d}t = \boldsymbol{\xi}_T(\boldsymbol{x}), \quad \boldsymbol{\xi}_T \in \boldsymbol{L}^2(\Omega),$$

assuming that $g_p \in \mathrm{C}([0,T])$ *satisfies either*

$$g_p(t) > 0 \quad \forall t \in [0,T]$$

or

$$g_p(t) < 0 \quad \forall t \in [0, T].$$

It is assumed that

- *The solution satisfies the properties of Lemma 1.1(i);*
- *The kernel k is strongly positive definite.*

Now, we state a result concerning the recovery of a space-dependent heat source; see [5, Theorem 3.1].

Lemma 2.5 (Problem 2.1: answer IV) *Consider the system*

$$\begin{cases} \rho\partial_{tt}\boldsymbol{u} - \mu\Delta\boldsymbol{u} - (\lambda+\mu)\nabla(\nabla\cdot\boldsymbol{u}) + \gamma\nabla\theta = \boldsymbol{0} & \text{in } Q_T; \\ \rho C_s\partial_t\theta - \kappa\Delta\theta - k*\Delta\theta + T_0\gamma\nabla\cdot\partial_t\boldsymbol{u} = g_h(t)f_h(\boldsymbol{x}) & \text{in } Q_T; \\ \boldsymbol{u} = \boldsymbol{0}, \quad \theta = 0 & \text{on } \Sigma_T; \\ \boldsymbol{u}(\cdot,0) = \partial_t\boldsymbol{u}(\cdot,0) = \boldsymbol{0}, \quad \theta(\cdot,0) = 0 & \text{in } \Omega. \end{cases}$$

It is possible to recover $f_h(\boldsymbol{x})$ uniquely for all types of thermoelasticity from

$$\int_0^T \theta(\boldsymbol{x},t)\,\mathrm{d}t = \zeta_T(\boldsymbol{x}), \quad \zeta_T \in \mathrm{L}^2(\Omega),$$

assuming that $g_h \in \mathrm{C}^1([0,T])$ satisfies either

$$g_h(t) > 0 \quad \text{and} \quad g_h'(t) \geq 0, \quad \forall t \in [0,T]$$

or

$$g_h(t) < 0 \quad \text{and} \quad g_h'(t) \leq 0, \quad \forall t \in [0,T].$$

It is assumed that:

- *The solution satisfies the properties of Lemma 1.1(ii);*
- *The kernel k is strongly positive definite.*

The uniqueness of a solution to this problem when considering the final in time measurement $\theta(\cdot, T)$ is still unanswered. The results of Lemmas 2.3, 2.4, and 2.5 stay valid when a homogeneous Neumann boundary condition for the temperature is considered. Moreover, in [5], counterexamples for uniqueness are constructed, showing the necessity of the sign condition on $g_{\mathbf{p}}$ and g_h in these lemmas. Finally, we mention a result concerning the simultaneous determination of a space-dependent load and heat source; see [5, Theorem 4.1]. It is assumed that $g_{\mathbf{p}}$ and g_h are proportional.

Lemma 2.6 (Problem 2.1: answer V) *Consider the system*

$$\begin{cases} \rho\partial_{tt}\boldsymbol{u} - \mu\Delta\boldsymbol{u} - (\lambda+\mu)\nabla(\nabla\cdot\boldsymbol{u}) + \gamma\nabla\theta = g_{\boldsymbol{p}}(t)\boldsymbol{f_p}(\boldsymbol{x}) & \text{in } Q_T; \\ \rho C_s\partial_t\theta - \kappa\Delta\theta - k*\Delta\theta + T_0\gamma\nabla\cdot\partial_t\boldsymbol{u} = g_h(t)f_h(\boldsymbol{x}) & \text{in } Q_T; \\ \boldsymbol{u} = \boldsymbol{0}, \quad \theta = 0 & \text{on } \Sigma_T; \\ \boldsymbol{u}(\cdot,0) = \partial_t\boldsymbol{u}(\cdot,0) = \boldsymbol{0}, \quad \theta(\cdot,0) = 0 & \text{in } \Omega. \end{cases}$$

It is possible to recover $f_{\boldsymbol{p}}(\boldsymbol{x})$ and $f_h(\boldsymbol{x})$ uniquely for all types of thermoelasticity from

$$\boldsymbol{u}(\boldsymbol{x},T) = \boldsymbol{\xi}_T(\boldsymbol{x}), \quad \boldsymbol{\xi}_T \in \boldsymbol{L}^2(\Omega), \quad \text{and} \quad \int_0^T \theta(\boldsymbol{x},t)\,\mathrm{d}t = \zeta_T(\boldsymbol{x}), \quad \zeta_T \in \mathrm{L}^2(\Omega),$$

assuming that

$$g_{\boldsymbol{p}} = c g_h \quad \text{for some } c > 0,$$

and either

$$g_h(t) > 0 \quad \text{and} \quad g_h'(t) > 0, \quad \forall t \in [0,T]$$

or

$$g_h(t) < 0 \quad \text{and} \quad g_h'(t) < 0, \quad \forall t \in [0,T].$$

It is assumed that:

- *The solution satisfies the properties of Lemma 1.1(ii);*
- *The kernel k is strongly positive definite.*

3 Identification of the Time-Dependent Source g_h

In this section, the inverse source problem under consideration reads as follows.

Problem 3.1 Can the heat source $h(t)$ be theoretically and numerically determined from the knowledge of the additional integral measurement

$$m(t) := \int_\Omega \omega(\mathbf{x})\theta(\mathbf{x},t)\,\mathrm{d}\mathbf{x}, \quad t \in [0,T],$$

in the following isotropic thermoelasticity system of type III:

$$\begin{cases} \rho\partial_{tt}\mathbf{u} - \mu\Delta\mathbf{u} - (\lambda+\mu)\nabla(\nabla\cdot\mathbf{u}) + \gamma\nabla\theta = \mathbf{p} & \text{in } Q_T; \\ \rho C_s\partial_t\theta - \kappa\Delta\theta - k * \Delta\theta + T_0\gamma\nabla\cdot\partial_t\mathbf{u} = g_h(t)f_h(\mathbf{x}) + \tilde{h} & \text{in } Q_T, \\ \text{boundary conditions to be specified}, \end{cases}$$

together with the initial conditions

$$\mathbf{u}(\mathbf{x},0) = \overline{\mathbf{u}}_0(\mathbf{x}), \quad \partial_t\mathbf{u}(\mathbf{x},0) = \overline{\mathbf{u}}_1(\mathbf{x}), \quad \theta(\mathbf{x},0) = \overline{\theta}_0(\mathbf{x}), \quad \mathbf{x}\in\Omega,$$

for all types of thermoelasticity?

Taking advantage of the boundary conditions and the one-dimensional setting, in [3], an existence and uniqueness result for exact data has been obtained by employing Rothe's method. Using the measurement, the unknown g_h can be expressed in terms of u and θ. Assuming $\int_0^L f_h \neq 0$, we have that

$$g_h(t) = \frac{\rho C_s m'(t) - \kappa\int_0^L \theta_{xx}(t) - \left(k * \int_0^L \theta_{xx}\right)(t) + \gamma T_0\int_0^L u_{xt}(t) - \int_0^L \tilde{h}(t)}{\int_0^L f_h}.$$

Substituting this expression into the variational formulation (in the sense of Lemma 1.1(ii), see [3, Eq. (4–5)]), the inverse problem has been reformulated as a coupled direct problem with unknowns u and θ. This approach also contains a decoupled scheme for computations; see [3, Eq. (9–12)] for more details.

Lemma 3.2 (Problem 3.1: answer I) *Let $d = 1$ and $\omega = 1$. Consider $u' = \theta = 0$ on Γ. Assume that $k \in \mathrm{C}([0,T])$, $f_h \in \mathrm{H}^1(\Omega)$ with $\int_0^L f_h \neq 0$, $\tilde{h} \in \mathrm{L}^2\left([0,T],\mathrm{L}^2(\Omega)\right)$, $p \in \mathrm{L}^2\left([0,T],\mathrm{H}^1(\Omega)\right)$, $\overline{u}_0 \in \mathrm{H}^2(\Omega)$, $\overline{u}_1 \in \mathrm{H}^1(\Omega)$ and $\overline{\theta}_0 \in \mathrm{H}^2(\Omega)\cap\mathrm{H}^1_0(\Omega)$. Then, there exists a unique weak solution $\{u,\theta,h\}$ to Problem 3.1 satisfying:*

- $u \in \mathrm{L}^2\left((0,T),\mathrm{H}^2(\Omega)\right)\cap\mathrm{C}\left([0,T],\mathrm{H}^1(\Omega)\right)$, $u_t \in \mathrm{L}^2\left((0,T),\mathrm{H}^1(\Omega)\right)$, $u_{tt} \in \mathrm{L}^2\left((0,T),\mathrm{L}^2(\Omega)\right)$;
- $\theta \in \mathrm{L}^2\left((0,T),\mathrm{H}^2(\Omega)\cap\mathrm{H}^1_0(\Omega)\right)\cap\mathrm{C}\left([0,T],\mathrm{L}^2(\Omega)\right)$, $\theta_t \in \mathrm{L}^2\left((0,T),\mathrm{L}^2(\Omega)\right)$;
- $h \in \mathrm{L}^2(0,T)$.

The approach followed in the case of a Dirichlet boundary condition for θ does not seem possible to be extended to the multi-dimensional case since the term $\int_\Omega \Delta\theta$ in the expression for g_h cannot be controlled. In [6], we established the global-in-time existence by transferring derivatives from $\mathbf{u}$ and θ to the weight function ω in the measurement. We have considered the homogeneous Dirichlet boundary condition for $\mathbf{u}$ and have investigated the influence of the choice of the thermal boundary condition for θ on the analysis by considering the Dirichlet and Neumann boundary condition, respectively. If $-\kappa\nabla\theta\cdot\boldsymbol{\nu}=\tilde{g}$ on Γ, then

$$g_h(t)=\Big[\rho C_s m'(t)+\kappa\left(\nabla\theta(t),\nabla\omega\right)+\left(\tilde{g}(t),\omega\right)_\Gamma+\left((k*\nabla\theta)(t),\nabla\omega\right)$$

$$+\frac{1}{\kappa}\left((k*\tilde{g})(t),\omega\right)_\Gamma-\gamma T_0\left(\partial_t\mathbf{u}(t),\nabla\omega\right)-\left(\tilde{h}(t),\omega\right)\Big]/(f_h,\omega),\qquad \omega\in\mathrm{H}^1(\Omega),$$

where

$$(f_h,\omega)=\int_\Omega f_h(\mathbf{x})\omega(\mathbf{x})\,\mathrm{d}\mathbf{x}\neq 0.$$

Moreover, if $\theta=0$ on Γ, then the expression for g_h is given by

$$g_h(t)=\big[\rho C_s m'(t)+\kappa\left(\nabla\theta(t),\nabla\omega\right)+\left((k*\nabla\theta)(t),\nabla\omega\right)$$

$$-\gamma T_0\left(\partial_t\mathbf{u}(t),\nabla\omega\right)-\left(\tilde{h}(t),\omega\right)\big]/(f_h,\omega),\qquad \omega\in\mathrm{H}^1_0(\Omega).$$

Note the different conditions on the weight function in both expression for g_h. The assumption that $\omega\in\mathrm{H}^1_0(\Omega)$ for the Dirichlet boundary condition is to get rid of the boundary terms appearing in the expression of g_h after application of the Green theorem. The existence and uniqueness of a weak solution for exact data (in the sense of Lemma 1.1(i), see [6, Eq. (5-6) and (8-9)]) has been proved (including a numerical algorithm) by Rothe's method, and the result can be found in the following lemmas; see [6, Theorem 3.1 and 4.1].

Lemma 3.3 (Problem 3.1: answer II) *Let $\boldsymbol{u}=\boldsymbol{0}$ and $-\kappa\nabla\theta\cdot\boldsymbol{\nu}=\tilde{g}$ on Γ. Assume that $\overline{\boldsymbol{u}}_0\in\boldsymbol{H}^1_0(\Omega)$, $\overline{\boldsymbol{u}}_1\in\boldsymbol{L}^2(\Omega)$, $\overline{\theta}_0\in\mathrm{H}^1(\Omega)$, $\omega\in\mathrm{H}^1(\Omega)$, $f\in\mathrm{L}^2(\Omega)$ with $\int_\Omega f(\boldsymbol{x})\omega(\boldsymbol{x})\,\mathrm{d}\boldsymbol{x}\neq 0$, $k\in\mathrm{C}([0,T])$, $m\in\mathrm{C}^1([0,T])$, $\tilde{h}\in\mathrm{L}^\infty\left([0,T],\mathrm{H}^1(\Omega)\right)$, $\boldsymbol{p}\in\mathrm{L}^\infty\left([0,T],\boldsymbol{H}^1(\Omega)\right)$ and $g\in\mathrm{L}^\infty\left([0,T],\mathrm{L}^2(\Gamma)\right)$. Then, there exists a unique weak solution $\{\boldsymbol{u},\theta,h\}$ to Problem 3.1 such that:*

- $\boldsymbol{u}\in\mathrm{C}\left([0,T],\boldsymbol{L}^2(\Omega)\right)\cap\mathrm{L}^2\left((0,T),\boldsymbol{H}^1_0(\Omega)\right)$, $\partial_t\boldsymbol{u}\in\mathrm{L}^2\left((0,T),\boldsymbol{L}^2(\Omega)\right)\cap\mathrm{C}\left([0,T],\boldsymbol{H}^1_0(\Omega)^*\right)$, $\partial_{tt}\boldsymbol{u}\in\mathrm{L}^2\left((0,T),\boldsymbol{H}^1_0(\Omega)^*\right)$;
- $\theta\in\mathrm{C}\left([0,T],\boldsymbol{L}^2(\Omega)\right)\cap\mathrm{L}^2\left((0,T),\mathrm{H}^1(\Omega)\right)$, $\partial_t\theta\in\mathrm{L}^2\left((0,T),\mathrm{H}^1(\Omega)^*\right)$;
- $h\in\mathrm{L}^2(0,T)$.

Lemma 3.4 (Problem 3.1: answer III) *Let $\boldsymbol{u}=\boldsymbol{0}$ and $\theta=0$ on Γ. Assume that $\overline{\boldsymbol{u}}_0\in\boldsymbol{H}^1_0(\Omega)$, $\overline{\boldsymbol{u}}_1\in\boldsymbol{L}^2(\Omega)$, $\overline{\theta}_0\in\mathrm{H}^1_0(\Omega)$, $\omega\in\mathrm{H}^1_0(\Omega)$, $f\in\mathrm{L}^2(\Omega)$ with $\int_\Omega f(\boldsymbol{x})\omega(\boldsymbol{x})\,\mathrm{d}\boldsymbol{x}\neq 0$, $k\in\mathrm{C}([0,T])$, $m\in\mathrm{C}^1([0,T])$, $\tilde{h}\in\mathrm{L}^\infty\left([0,T],\mathrm{H}^1(\Omega)\right)$ and $\boldsymbol{p}\in\mathrm{L}^\infty\left([0,T],\boldsymbol{H}^1(\Omega)\right)$. Then, there exists a unique weak solution $\{\boldsymbol{u},\theta,h\}$ to Problem 3.1 such that:*

- $\boldsymbol{u} \in \mathrm{C}\left([0,T], \boldsymbol{L}^2(\Omega)\right) \cap \mathrm{L}^2\left((0,T), \boldsymbol{H}_0^1(\Omega)\right)$, $\partial_t \boldsymbol{u} \in \mathrm{L}^2\left((0,T), \boldsymbol{L}^2(\Omega)\right) \cap \mathrm{C}\left([0,T], \boldsymbol{H}_0^1(\Omega)^*\right)$, $\partial_{tt}\boldsymbol{u} \in \mathrm{L}^2\left((0,T), \boldsymbol{H}_0^1(\Omega)^*\right)$;
- $\theta \in \mathrm{C}\left([0,T], \boldsymbol{L}^2(\Omega)\right) \cap \mathrm{L}^2\left((0,T), \mathrm{H}_0^1(\Omega)\right)$, $\partial_t\theta \in \mathrm{L}^2\left((0,T), \mathrm{H}_0^1(\Omega)^*\right)$;
- $h \in \mathrm{L}^2(0,T)$.

Acknowledgments K. Van Bockstal was supported by the Methusalem programme of Ghent University Special Research Fund (BOF) (Grant Number 01M01021).

References

1. M.A. Biot, Thermoelasticity and irreversible thermodynamics. J. Appl. Phys. **27**(3), 240–253 (1956)
2. K. Van Bockstal, M. Slodička, Recovery of a space-dependent vector source in thermoelastic systems. Inverse Prob. Sci. Eng. **23**(6), 956–968 (2015)
3. K. Van Bockstal, M. Slodička, Recovery of a time-dependent heat source in one-dimensional thermoelasticity of type-III. Inverse Prob. Sci. Eng. **25**(5), 749–770 (2017)
4. K. Van Bockstal, L. Marin, Recovery of a space-dependent vector source in anisotropic thermoelastic systems. Comput. Methods Appl. Mech. Eng. **321**, 269–293 (2017)
5. F. Maes, K. Van Bockstal, Uniqueness for inverse source problems of determining a space-dependent source in thermoelastic systems. J. Inverse Ill-Posed Probl. **30**(6), 845–856 (2022)
6. K. Van Bockstal, L. Marin, Finite element method for the reconstruction of a time-dependent heat source in isotropic thermoelasticity systems of type-III. Z. Angew. Math. Phys. **73**(3), 113 (2022)
7. M. Bellassoued, M. Yamamoto, Carleman estimates and an inverse heat source problem for the thermoelasticity system. Inverse Prob. **27**(1), 015006 (2011)
8. B. Wu, J. Liu, Determination of an unknown source for a thermoelastic system with a memory effect. Inverse Prob. **28**(9), 095012 (2012)
9. J.A. Nohel, D.F. Shea, Frequency domain methods for Volterra equations. Adv. Math. **22**(3), 278–304 (1976)
10. T. Johansson, D. Lesnic, Determination of a spacewise dependent heat source. J. Comput. Appl. Math. **209**(1), 66–80 (2007)
11. V. Isakov, *Inverse Source Problems*. (American Mathematical Society, Providence, RI, 1990)

A Direct Method of Moving Planes for Logarithmic Schrödinger Operator

Rong Zhang

Abstract In this paper, we study the radial symmetry and monotonicity of nonnegative solutions to nonlinear equations involving the logarithmic Schrödinger operator $(\mathcal{I}-\Delta)^{\log}$ corresponding to the logarithmic symbol $\log(1+|\xi|^2)$, which is a singular integral operator given by

$$(\mathcal{I}-\Delta)^{\log}u(x) = c_N P.V. \int_{\mathbb{R}^N} \frac{u(x)-u(y)}{|x-y|^N}\kappa(|x-y|)dy,$$

where $c_N = \pi^{-\frac{N}{2}}$, $\kappa(r) = 2^{1-\frac{N}{2}} r^{\frac{N}{2}} \mathcal{K}_{\frac{N}{2}}(r)$, and $\mathcal{K}_\nu$ is the modified Bessel function of the second kind with index ν. The proof hinges on a direct method of moving planes for the logarithmic Schrödinger operator. For a more detailed analysis and for the proofs of the announced results, we refer to (Zhang R, Kumar V, Ruzhansky M, A direct method of moving planes for logarithmic schrödinger operator. arXiv:2210.09811).

2000 Mathematics Subject Classification 35R11, 35B06, 35B50, 35B51, 35D30

1 Introduction

The study of Schrödinger equations received a great deal of attention from researchers in the past decades because of its vast applications in several areas of mathematics and mathematical physics. In particular, Schrödinger equations arise in quantum field theory and in the Hartree-Fock theory (see [5, 17, 18, 21]).

R. Zhang (✉)
HLM, Academy of Mathematics and Systems Science, Chinese Academy of Sciences, Beijing, China

Department of Mathematics: Analysis, Logic and Discrete Mathematics, Ghent University, Ghent, Belgium
e-mail: rongzhangnnu@163.com

M. Ruzhansky, B. Torebek (eds.), *Extended Abstracts MWCAPDE 2023*,
Research Perspectives Ghent Analysis and PDE Center 1,
https://doi.org/10.1007/978-3-031-41665-1_25

Recently, there is a surge of interest to investigate integrodifferential operators of order close to zero and associated linear and nonlinear integrodifferential equations (see [1, 4, 14, 16, 19, 20]). In particular, the logarithmic Laplacian and the logarithmic Schrödinger operator are two interesting examples of such class of operators. The logarithmic Schrödinger operator $(\mathcal{I}-\Delta)^{\log}$ (see [10]) and the logarithmic Laplacian L_{Δ} (see [1, 12]) have the similar behavior locally concerning to the singularity of kernels, but the logarithmic Schrödinger operator eliminates the integrability problem of the logarithmic Laplacian at infinity. To define the logarithmic Schrödinger operator, let us begin with the following observation:

$$\lim_{s\to 0^+}(\mathcal{I}-\Delta)^s u(x)=u(x),\ \text{ for } u\in C^2(\mathbb{R}^N), \tag{1.1}$$

where for $s\in(0,1)$, the operator $(\mathcal{I}-\Delta)^s$ stands for the relativistic Schrödinger operator, for sufficiently regular function $u:\mathbb{R}^N\to\mathbb{R}$, which can be represented via hypersingular integral (1.1) (see [8])

$$(\mathcal{I}-\Delta)^s u(x)=u(x)+c_{N,s}\lim_{\epsilon\to 0^+}\int_{\mathbb{R}^N\setminus B_\epsilon(0)}\frac{u(x)-u(y)}{|x-y|^{N+2s}}\varpi_s(|x-y|)dy, \tag{1.2}$$

where $c_{N,s}=\frac{\pi^{-\frac{N}{2}}4^s}{\Gamma(-s)}$ is a normalization constant and the function ϖ_s is given by

$$\varpi_s(r)=2^{1-\frac{N+2s}{2}}r^{\frac{N+2s}{2}}\mathcal{K}_{\frac{N+2s}{2}}(r)=\int_0^{+\infty}t^{-1+\frac{N+2s}{2}}e^{-t-\frac{r^2}{4t}}dt. \tag{1.3}$$

Here the function $\mathcal{K}_\nu$ is the modified Bessel function of the second kind with index $\nu>0$ (see [6, 7, 9, 10, 13]).

It follows from (1.1) that one may expect a Taylor expansion with respect to parameter s of the operator $(\mathcal{I}-\Delta)^s$ near zero for $u\in C^2(\mathbb{R}^N)$ and $x\in\mathbb{R}^N$ as

$$(\mathcal{I}-\Delta)^s u(x)=u(x)+s(\mathcal{I}-\Delta)^{\log}u(x)+o(s),\quad \text{as } s\to 0^+. \tag{1.4}$$

The logarithmic Schrödinger operator $(\mathcal{I}-\Delta)^{\log}$ appears as the first-order term in the above expansion.

Now we study the integrodifferential operator $(\mathcal{I}-\Delta)^{\log}$ corresponding to the logarithmic symbol $\log(1+|\xi|^2)$, which is a singular integral operator given by

$$(\mathcal{I}-\Delta)^{\log}u(x)=c_N P.V.\int_{\mathbb{R}^N}\frac{u(x)-u(y)}{|x-y|^N}\kappa(|x-y|)dy, \tag{1.5}$$

where $c_N=\pi^{-\frac{N}{2}}$, $P.V.$ stands for the Cauchy principal value of the integral, $\kappa(r)=2^{1-\frac{N}{2}}r^{\frac{N}{2}}\mathcal{K}_{\frac{N}{2}}(r)$ and $\mathcal{K}_\nu$ is the modified Bessel function of second kind with index ν. Using the expression (1.5), one can define $(\mathcal{I}-\Delta)^{\log}$ for a quite large class of functions u. To illustrate this, define the space $\mathcal{L}_0(\mathbb{R}^N)$ as the space of locally integrable functions $u : \mathbb{R}^N \to \mathbb{R}$ such that

$$\|u\|_{\mathcal{L}_0(\mathbb{R}^N)} := \int_{\mathbb{R}^N} \frac{|u(x)|e^{-|x|}}{(1+|x|)^{\frac{N+1}{2}}} dx < +\infty.$$

Then, it was proved by [10, Proposition 2.1] that for $u \in \mathcal{L}_0(\mathbb{R}^N) \cap L^\infty(\mathbb{R}^N)$ which is also Dini continuous at some $x \in \mathbb{R}^N$, the quantity $[(\mathcal{I} - \Delta)^{\log} u](x)$ is well defined by formula (1.5). Let us recall the definition of Dini continuity. Let U be a measurable subset of $\mathbb{R}^N$ and let $u : U \to \mathbb{R}$ be a measurable function. The modulus of continuity $\Psi_{u,x,U} : (0, +\infty) \to [0, +\infty)$ of u at a point $x \in U$ is defined by

$$\Psi_{u,x,U}(r) := \sup_{y \in U, |x-y| \leq r} |u(x) - u(y)|.$$

We call the function u Dini continuous at x if

$$\int_0^1 \frac{\Psi_{u,x,U}(r)}{r} dr < \infty.$$

Using the generalized direct method of moving planes, in this note we obtain the radial symmetry and monotonicity of nonnegative solutions for the nonlinear equations involving the logarithmic Schrödinger operator (see Theorem 1.1), namely, we consider the nonlinear Schrödinger equation (For a more detailed analysis and for the proofs of the announced results, we refer to [24])

$$(\mathcal{I} - \Delta)^{\log} u(x) + mu(x) = u^p(x), \quad x \in \mathbb{R}^N, \tag{1.6}$$

with $u(x) \geq 0$ for all $x \in \mathbb{R}^N$.

The following result present symmetry and monotonicity properties of the Schrödinger equation (1.6).

Theorem 1.1 *Let $u \in \mathcal{L}_0(\mathbb{R}^N)$ be a nonnegative Dini continuous solution of* (1.6) *with $m > 0$ and $1 < p < \infty$. If*

$$\lim_{|x| \to \infty} u(x) = a < \left(\frac{m}{p}\right)^{\frac{1}{p-1}}, \tag{1.7}$$

then u must be radially symmetric and monotone decreasing about some point in $\mathbb{R}^N$.

2 Key Ingredients for the Method of Moving Planes

This section is devoted to developing basic and key results needed to apply the method of moving planes for establishing the proof of our main result in the next section. We first present some basic notation and nomenclatures which will be beneficial for the rest of the paper.

Choose an arbitrary direction, say, the x_1-direction. For arbitrary $\lambda \in \mathbb{R}$, let

$$T_\lambda = \{x \in \mathbb{R}^N \mid x_1 = \lambda\}$$

be the moving plane, and let

$$\Sigma_\lambda = \{x \in \mathbb{R}^N \mid x_1 < \lambda\}$$

be the region to the left of the plane T_λ, and

$$x^\lambda = (2\lambda - x_1, x_2, \cdots, x_N)$$

be the reflection of x about the plane T_λ.

By denoting $u(x^\lambda) := u_\lambda(x)$, we define

$$\omega_\lambda(x) := u_\lambda(x) - u(x), \quad x \in \Sigma_\lambda,$$

to compare the values of $u(x)$ and $u_\lambda(x)$.

The following results on the strong maximum principle for the operator $(\mathcal{I} - \Delta)^{\log}$ can be deduced from [15, Theorem 1.1] (see also [11] and [10, Theorem 6.1]).

Lemma 2.1 *(Strong maximum principle) Let $\Omega \subset \mathbb{R}^N$ be a domain, and let $u \in \mathcal{L}_0(\mathbb{R}^N)$ be a continuous function on $\bar{\Omega}$ satisfying*

$$\begin{cases} (\mathcal{I} - \Delta)^{\log} u(x) \geq 0, & in\ \Omega, \\ u(x) \geq 0, & in\ \mathbb{R}^N \backslash \Omega, \end{cases} \tag{2.1}$$

then $u > 0$ in Ω or $u = 0$ a.e. in $\mathbb{R}^N$.

Theorem 2.2 (Maximum principle for antisymmetric functions) *Let Ω be a bounded domain in Σ_λ. Assume that $\omega_\lambda \in \mathcal{L}_0(\mathbb{R}^N) \cap L^\infty(\mathbb{R}^N)$ is Dini continuous on Ω and is lower semi-continuous on $\bar{\Omega}$. If*

$$\begin{cases} (\mathcal{I} - \Delta)^{\log} \omega_\lambda(x) \geq 0, & in\ \Omega, \\ \omega_\lambda(x) \geq 0, & in\ \Sigma_\lambda \backslash \Omega, \\ \omega_\lambda(x^\lambda) = -\omega(x), & in\ \Sigma_\lambda, \end{cases} \tag{2.2}$$

then

$$\omega_\lambda \geq 0 \quad in\ \Omega. \tag{2.3}$$

Furthermore, if $\omega_\lambda(x) = 0$ at some point in Ω, then we have

$$\omega_\lambda = 0,\ a.e.\ in\ \mathbb{R}^N. \tag{2.4}$$

These conclusions hold for unbounded region Ω if we further assume that

$$\liminf_{|x|\to\infty} \omega_\lambda(x) \geq 0.$$

Proof If ω_λ is not nonnegative on Ω, then the lower semi-continuity of ω_λ on $\bar{\Omega}$ implies that there exists a $x^o \in \bar{\Omega}$ such that

$$\omega_\lambda(x^o) := \min_{\bar{\Omega}} \omega_\lambda(x) < 0.$$

One can further deduce from (2.2) that x^o is in the interior of Ω. It follows that

$$\begin{aligned}(\mathcal{I}-\Delta)^{\log}\omega_\lambda(x^o) =& c_N P.V. \int_{\mathbb{R}^N} \frac{\omega_\lambda(x^o)-\omega_\lambda(y)}{|x^o-y|^N}\kappa(|x^o-y|)dy \\ =& c_N P.V. \Bigg(\int_{\Sigma_\lambda} \frac{\omega_\lambda(x^o)-\omega_\lambda(y)}{|x^o-y|^N}\kappa(|x^o-y|)dy \\ &+ \int_{\Sigma_\lambda} \frac{\omega_\lambda(x^o)-\omega_\lambda(y^\lambda)}{|x^o-y^\lambda|^N}\kappa(|x^o-y^\lambda|)dy \Bigg).\end{aligned} \tag{2.5}$$

Since $|x^o - y| \leq |x^o - y^\lambda|$ we have $\frac{1}{|x^o-y|} \geq \frac{1}{|x^o-y^\lambda|}$ and $\kappa(|x^o-y|) \geq \kappa(|x^o-y^\lambda|)$ as κ is a decreasing function, and, therefore,

$$\frac{\omega_\lambda(x^o)-\omega_\lambda(y)}{|x^o-y|^N}\kappa(|x^o-y|) \leq \frac{\omega_\lambda(x^o)-\omega_\lambda(y)}{|x^o-y^\lambda|^N}\kappa(|x^o-y^\lambda|),$$

since $\omega_\lambda(x^o) - \omega_\lambda(y) \leq 0$. Thus, we obtain from (2.5) that

$$\begin{aligned}(\mathcal{I}-\Delta)^{\log}\omega_\lambda(x^o) \leq& c_N P.V. \int_{\Sigma_\lambda} \left(\frac{\omega_\lambda(x^o)-\omega_\lambda(y)}{|x^o-y^\lambda|^N} + \frac{\omega_\lambda(x^o)+\omega_\lambda(y)}{|x^o-y^\lambda|^N} \right) \\ &\times \kappa(|x^o-y^\lambda|)dy \\ =& c_N P.V. \int_{\Sigma_\lambda} \frac{2\omega_\lambda(x^o)}{|x^o-y^\lambda|^N}\kappa(|x^o-y^\lambda|)dy < 0,\end{aligned} \tag{2.6}$$

which contradicts (2.2). Therefore, our assumption is wrong, and consequently, we have $\omega_\lambda(x) \geq 0$ in Ω.

Now we have proved that $\omega_\lambda(x) \geq 0$ in Ω. If there is some point $\tilde{x} \in \Omega$ such that $\omega_\lambda(\tilde{x}) = 0$, then from Lemma 2.1, we derive immediately $\omega_\lambda = 0$ a.e. in $\mathbb{R}^N$.

For unbounded domain Ω, the condition

$$\liminf_{|x|\to\infty} \omega_\lambda(x) \geq 0,$$

ensures that the negative minimum of ω_λ must be attained at some point x^o; then we can derive the same contradiction as above.

This completes the proof of Theorem 2.2. □

Theorem 2.3 (Decay at infinity) *Let Ω be an unbounded domain in Σ_λ. Suppose that a Dini continuous $\omega_\lambda \in \mathcal{L}_0(\mathbb{R}^N) \cap L^\infty(\mathbb{R}^N)$ is a solution to*

$$\begin{cases} (\mathcal{I} - \Delta)^{\log}\omega_\lambda(x) + c(x)\omega_\lambda(x) \geq 0, & in\ \Omega, \\ \omega_\lambda(x) \geq 0, & in\ \Sigma_\lambda \backslash \Omega, \\ \omega_\lambda(x^\lambda) = -\omega(x), & in\ \Sigma_\lambda, \end{cases} \tag{2.7}$$

with the measurable function $c(x)$ such that

$$\limsup_{|x|\to\infty} c(x) \geq 0. \tag{2.8}$$

Then there exists a constant $R_o > 0$, such that if

$$\omega_\lambda(x^o) = \min_{\Omega} \omega_\lambda(x) < 0, \tag{2.9}$$

then

$$|x^o| \leq R_o. \tag{2.10}$$

Proof We prove the assertion by contradiction. Suppose that (2.10) is false, then by (2.7) and (2.9), we have

$$\omega_\lambda(x^o) = \min_{\Sigma_\lambda} \omega_\lambda(x) < 0.$$

After a direct calculation, we obtain

$$
\begin{aligned}
(\mathcal{I}-\Delta)^{\log}\omega_\lambda(x^o) =& c_N P.V.\int_{\mathbb{R}^N}\frac{\omega_\lambda(x^o)-\omega_\lambda(y)}{|x^o-y|^N}\kappa(|x^o-y|)dy\\
=& c_N P.V.\int_{\Sigma_\lambda}\Bigg(\frac{\omega_\lambda(x^o)-\omega_\lambda(y)}{|x^o-y|^N}\kappa(|x^o-y|)\\
&+\frac{\omega_\lambda(x^o)-\omega_\lambda(y^\lambda)}{|x^o-y^\lambda|^N}\kappa(|x^o-y^\lambda|)\Bigg)dy\\
\le& c_N P.V.\int_{\Sigma_\lambda}\left(\frac{\omega_\lambda(x^o)-\omega_\lambda(y)}{|x^o-y^\lambda|^N}+\frac{\omega_\lambda(x^o)+\omega_\lambda(y)}{|x^o-y^\lambda|^N}\right)\\
&\times\kappa(|x^o-y^\lambda|)dy\\
=& c_N P.V.\int_{\Sigma_\lambda}\frac{2\omega_\lambda(x^o)}{|x^o-y^\lambda|^N}\kappa(|x^o-y^\lambda|)dy<0.
\end{aligned}
\tag{2.11}
$$

Now, we fix λ and when $|x^o|\ge\lambda$, we have $B_{|x^o|}(\breve{x})\subset\tilde{\Sigma}_\lambda:=\mathbb{R}^N\backslash\Sigma_\lambda$ with $\breve{x}=(3|x^o|+x_1^o,(x^o)')$. So, for $y\in\Sigma_\lambda$, we can deduce that $|x^o-y|\le|x^o-\breve{x}|+|\breve{x}-y|\le|x^o|+3|x^o|=|4x^o|$ which together with the fact that κ is a decreasing function implies that

$$
\frac{\kappa(|x^o-y|)}{|x^o-y|}\ge\frac{\kappa(|4x^o|)}{|4x^o|}.
$$

Thus, if $|x^o|\ge\frac{R}{4}$ with sufficiently large R then from $\kappa(r)=2^{1-\frac{N}{2}}r^{\frac{N}{2}}\mathcal{K}_{\frac{N}{2}}(r)$, we have

$$
\begin{aligned}
\int_{\Sigma_\lambda}\frac{1}{|x^o-y^\lambda|^N}\kappa(|x^o-y^\lambda|)dy=\int_{\tilde{\Sigma}_\lambda}\frac{\kappa(|x^o-y|)}{|x^o-y|^N}dy\ge&\int_{B_{|x^o|}(\breve{x})}\frac{\kappa(|4x^o|)}{|4x^o|^N}dy\\
\ge&\int_{B_{|x^o|}(\breve{x})}\frac{2^{1-\frac{N}{2}}\mathcal{K}_{\frac{N}{2}}(|4x^o|)}{|4x^o|^{\frac{N}{2}}}dy\\
\ge&\frac{c_\infty\omega_N}{2^{\frac{3N}{2}}|x^o|^{\frac{1+N}{2}}e^{4|x^o|}}:=\frac{C}{|x^o|^{\frac{1+N}{2}}e^{4|x^o|}},
\end{aligned}
\tag{2.12}
$$

where $C=c_\infty\omega_N2^{-\frac{3N}{2}}$ is a positive constant. It follows that

$$
0\le(\mathcal{I}-\Delta)^{\log}\omega_\lambda(x^o)+c(x^o)\omega_\lambda(x^o)\le\left(\frac{C}{|x^o|^{\frac{1+N}{2}}e^{4|x^o|}}+c(x^o)\right)\omega_\lambda(x^o),
$$

or equivalently,

$$\frac{C}{|x^o|^{\frac{1+N}{2}} e^{4|x^o|}} + c(x^o) \leq 0.$$

Now, if $|x^o|$ is sufficiently large, this would contradict (2.8). Therefore, (2.10) holds.

This completes the proof of Theorem 2.3. □

3 Proof of the Main Theorem

Proof of Theorem 1.1 Let T_λ, Σ_λ, x^λ and ω_λ be defined as in the previous section. Then at the points where $\omega_\lambda(x) < 0$, it is easy to verify that, for $\xi_\lambda(x) \in (u_\lambda(x), u(x))$, we have

$$\begin{aligned}(\mathcal{I} - \Delta)^{\log} \omega_\lambda(x) + m\omega_\lambda(x) &= u_\lambda^p(x) - u^p(x) \\ &= p\xi_\lambda^{p-1}(x)\omega_\lambda(x) \geq pu^{p-1}(x)\omega_\lambda(x),\end{aligned} \tag{3.1}$$

because $\omega_\lambda(x) < 0$, and $\xi_\lambda(x) < u(x)$.

Step 1. We will show that for sufficiently negative λ,

$$\omega_\lambda(x) \geq 0, \quad x \in \Sigma_\lambda. \tag{3.2}$$

First, from the assumption (1.7), for each fixed λ, $\lim_{|x|\to\infty} \omega_\lambda(x) = 0$. In fact, by (1.7), we have $\lim_{|x|\to\infty} u(x) = a$, and $\lim_{|x|\to\infty} u_\lambda(x) = a$ implying that $\lim_{|x|\to\infty} \omega_\lambda(x) = 0$. Thus, if (3.2) is false, then the negative minimum of ω_λ can be obtained at some point, say, x^o in Σ_λ, that is,

$$\omega_\lambda(x^o) = \min_{\Sigma_\lambda} \omega_\lambda(x) < 0.$$

Set $c(x) := m - pu^{p-1}(x)$ in (3.1), and then, the assumption (1.7) implies that $c \in L^\infty(\mathbb{R}^N)$ and

$$\lim_{|x|\to\infty} c(x) \geq 0.$$

Consequently, from Theorem 2.3 it follows that there exists $R_o > 0$ (independent of λ), such that

$$|x^o| \leq R_o. \tag{3.3}$$

Therefore, by choosing $\lambda < -R_o$ and consequently, $|x^\lambda| > R_0$ for $x \in \Sigma_\lambda$, we obtain by (3.3) that

$$\omega_\lambda(x) \geq 0, \quad x \in \Sigma_\lambda. \tag{3.4}$$

Step 2. Step 1 provides a starting point, from which we can now move the plane T_λ to the right as long as (3.2) holds to its limiting position. Define

$$\lambda_o := \sup\{\lambda \mid \omega_\mu(x) \geq 0, \ \forall\, x \in \Sigma_\mu, \ \forall \mu \leq \lambda\}.$$

By (3.3), we know $\lambda_o < \infty$.

Next, we will show via a contradiction argument that

$$\omega_{\lambda_o}(x) \equiv 0, \quad \forall\, x \in \Sigma_{\lambda_o}. \tag{3.5}$$

Suppose, on the contrary, that

$$\omega_{\lambda_o}(x) \geq 0, \text{ and } \omega_{\lambda_o}(x) \not\equiv 0, \text{ in } \Sigma_{\lambda_o}, \tag{3.6}$$

then we must have

$$\omega_{\lambda_o}(x) > 0, \quad \forall\, x \in \Sigma_{\lambda_o}. \tag{3.7}$$

In fact, if (3.7) is violated, then there exists a point $\hat{x} \in \Sigma_{\lambda_o}$ such that

$$\omega_{\lambda_o}(\hat{x}) = \min_{\Sigma_{\lambda_o}} \omega_{\lambda_o}(x) = 0.$$

It means that $u_{\lambda_o}(\hat{x}) = u(\hat{x})$. Then it follows from (3.1) that

$$(\mathcal{I} - \Delta)^{\log} \omega_{\lambda_o}(\hat{x}) = u_{\lambda_o}^p(\hat{x}) - u^p(\hat{x}) = u^p(\hat{x}) - u^p(\hat{x}) = 0.$$

Hence, Theorem 2.2 implies that $\omega_{\lambda_o}(\hat{x}) \equiv 0$ in Σ_{λ_o}, which contradicts (3.6). Thus (3.7) holds.

Now, we will show that the plane T_λ can be moved further right. More precisely, there exists an $\epsilon > 0$ such that, for any $\lambda \in [\lambda_o, \lambda_o + \epsilon)$, we have

$$\omega_\lambda(x) \geq 0, \quad x \in \Sigma_\lambda. \tag{3.8}$$

Once it is proved, this will contradict the definition of λ_o. Therefore, (3.5) must be valid.

Let us now prove (3.8). In fact, by (3.7), we have $\omega_{\lambda_o}(x) > 0, \ x \in \Sigma_{\lambda_o}$, which in turn implies that there is a constant $c_o > 0$ and $\delta > 0$ such that

$$\omega_{\lambda_o}(x) \geq c_o > 0, \quad x \in \overline{\Sigma_{\lambda_o - \delta} \cap B_{R_o}(0)}.$$

Since ω_λ is continuous with respect to λ, there exists an $\epsilon > 0$ such that for $\lambda \in [\lambda_o, \lambda_o + \epsilon)$, we have

$$\omega_\lambda(x) \geq 0, \quad x \in \Sigma_{\lambda_o - \delta} \cap B_{R_o}(0). \tag{3.9}$$

Moreover, combining (3.3) with (3.9), we deduce that $w_\lambda(x) \geq 0$ on $\Sigma_{\lambda_0 - \delta}$.

To proceed with the proof, we need the following small volume maximum principle ([10, Theorem 6.1 (iii)] and [15, Theorem 1.3]).

Lemma 3.1 *Let Ω be an open set of $\mathbb{R}^N$. Consider the following problem on Ω:*

$$\begin{cases} (\mathcal{I} - \Delta)^{\log} u(x) \geq c(x)u, & in\ \Omega, \\ u \geq 0, & in\ \mathbb{R}^N \backslash \Omega, \end{cases} \tag{3.10}$$

with $c \in L^\infty(\mathbb{R}^N)$.

Then, there exists $\delta > 0$ such that for every open set $\Omega \subset \mathbb{R}^N$ with $|\Omega| \leq \delta$ and any solution $u \in \mathcal{V}_\omega(\Omega)$ of (3.10) in Ω, where the space $\mathcal{V}_\omega(\Omega)$ is given in [10, Section 6], we have $u \geq 0$ in $\mathbb{R}^N$.

Consequently, according to Lemma 3.1 (by taking $\Omega = (\Sigma_\lambda \setminus \Sigma_{\lambda_o - \delta}) \cap B_{R_o}(0)$), we obtain that (3.8) holds.

The arbitrariness of the x_1-direction leads to the radial symmetry of $u(x)$ about some point in $\mathbb{R}^N$, and the monotonicity is a consequence of the fact that (3.4) holds.

This completes the proof of the Theorem 1.1. □

Acknowledgments RZ is supported by the China Scholarship Council (No.202106860018), the National Natural Science Foundation of China (Grant No. 11871278), and the National Natural Science Foundation of China (Grant No. 11571093).

References

1. H. Chen, T. Weth, The Dirichlet problem for the logarithmic Laplacian. Commun. Partial Differ. Equ. **44**(11), 1100–1139 (2019)
2. W. Chen, C. Li, Classification of solutions of some nonlinear elliptic equations. Duke Math. J. **63**, 615–622 (1991)
3. W. Chen, C. Li, Y. Li, A direct method of moving planes for the fractional Laplacian. Adv. Math. **308**, 404–437 (2017)
4. E. Correa, A. de Pablo, Nonlocal operators of order near zero. J. Math. Anal. Appl. **461**, 837–867 (2018)
5. T. Cazenave, *Semilinear Schrödinger Equations* (AMS/Courant Institute of Mathematical Sciences, New York University, 2003)
6. A. Elgart, B. Schlein, Mean field dynamics of boson stars. Commun. Pure Appl. Math. **60**, 500–545 (2007)
7. J. Fröhlich, B. Jonsson, E. Lenzmann, Boson stars as solitary waves. Commun. Math. Phys. **274**, 1–30 (2007)

8. M. Fall, V. Felli, Sharp essential self-adjointness of relativistic Schrödinger operators with a singular potential. J. Funct. Anal. **267**, 1851–1877 (2014)
9. M. Fall, V. Felli, Unique continuation properties for relativistic Schrödinger operators with a singular potential. Discrete Contin. Dyn. Syst. A **35**, 5827–5867 (2015)
10. P. Feulefack, The logarithmic Schrödinger operator and associated Dirichlet problems (2021). arXiv preprint arXiv:2112.08783
11. R.L. Frank, E. Lenzmann, On ground states for the L^2-critical boson star equation. Preprint (2009). arXiv:0910.2721
12. R.L. Frank, T. König, H. Tang, Classification of solutions of an equation related to a conformal log Sobolev inequality. Adv. Math. **375**, 107395 (2020)
13. Y. Guo, S. Peng, Symmetry and monotonicity of nonnegative solutions to pseudo-relativistic Choquard equations. Z. Angew. Math. Phys. **72**, 120 (2021)
14. V. Hernández-Santamará, A. Saldaña, Small order asymptotics for nonlinear fractional problems (2021). arXiv preprint arXiv:2108.00448
15. S. Jarohs, T. Weth, On the maximum principle for nonlocal operators. Math. Z. **293**(1), 81–111 (2019)
16. M. Kassmann, A. Mimica, Intrinsic scaling properties for nonlocal operators. J. Eur. Math. Soc. **19**(4), 983–1011 (2017)
17. G. Lieberman, *Second Order Parabolic Differential Equations* (World Scientific Publishing, Singapore, 1996)
18. W. Liu, E. Kengne, *Schrödinger Equations in Nonlinear Systems* (Springer, Singapore, 2019)
19. A. Laptev, T. Weth, Spectral properties of the logarithmic Laplacian. Anal. Math. Phys. **133**, 24 (2021)
20. A. Mimica, On harmonic functions of symmetric Lévy processes. Ann. l'IHP Probabilités statistiques **50**, 214–235 (2014)
21. C. Sulem, P. Sulem, *The Nonlinear Schrödinger Equation: Self-Focusing and Wave Collapse* (Springer, New York, 2007)
22. G. Wang, X. Ren, Z. Bai, W. Hou, Radial symmetry of standing waves for nonlinear fractional Laplacian Hardy-Schrödinger systems. Appl. Math. Lett. **96**, 131–137 (2019)
23. L. Zhang, X. Nie, A direct method of moving planes for the Logarithmic Laplacian. Appl. Math. Lett. **118**, 107141 (2021)
24. R. Zhang, V. Kumar, M. Ruzhansky, A direct method of moving planes for logarithmic schrödinger operator. arXiv:2210.09811.

The manufacturer's authorised representative in the EU is Springer Nature Customer Service Centre GmbH, Europaplatz 3, 69115 Heidelberg, Germany. If you have any concerns regarding our products, please contact ProductSafety@springernature.com

Printed and bound by CPI Group (UK) Ltd, Croydon, CR0 4YY
01/10/2026
02225851-0004